83384

THE
SATANIC
GASES

THE SATANIC GASES

Clearing the Air
about Global Warming

BY PATRICK J. MICHAELS AND
ROBERT C. BALLING, JR.

CATO
INSTITUTE
Washington, D.C.

Library of Congress Cataloging-in-Publication Data

Michaels, Patrick J.
   The satanic gases : clearing the air about global warming / by
Patrick J. Michaels and Robert C. Balling, Jr.
      p. cm.
   Includes bibliographical references and index.
   ISBN 1-882577-91-4. — ISBN 1-882577-92-2
   1. Global warming.   I. Balling, Robert C.   II. Title

QC981.8.G56 .M537   2000
363.738'74—dc21   00-022645

Printed in the United States of America.

CATO INSTITUTE
1000 Massachusetts Ave., N.W.
Washington, D.C. 20001

Contents

Acknowledgments

As first author, I had the opportunity to write this book because of a half-time appointment as senior fellow in environmental studies at Cato Institute. It is a remarkable place to work, as the Institute's influence is far out of proportion to its modest budget. Cato owes its success, and I owe this book, to Ed Crane, who runs the Institute.

Actually, he does not "run" it. Cato runs itself. Ed's genius is that he knows that the people he hires will not have to be told what to do, which is why Cato is so efficient. Cato also has a lovely building and, for folks who do not spend a lot of time in church, the classiest Christmas decorations in Washington. I could not imagine that working for someone else could be so much fun.

Jerry Taylor, director of natural resource studies at Cato, was responsible for bringing me to the Institute, and he has labored for days and nights over the various versions of this manuscript. I have been around environmental issues for decades and have never found anyone as well read as Jerry in the related science, economics, and policy. If I need to steal some weighty tome on these matters, I sneak up to his office. He has so many books along these lines that he has yet to notice the attrition. This book has benefited immeasurably from his efforts and his commitment to understanding the environmental issues we all confront.

James Ruin Lawrence, an intern at Cato, provided invaluable technical assistance. He is now completing his studies in Physics and Philosophy at Washington University. Should he apply for a position with your company you will do very well to hire him.

I also owe much of this book to the people at New Hope Environmental Services, a little company I started six years ago in New Hope, Virginia, dedicated to getting the facts out on a number of environmental issues, including global warming. New Hope also employs second author Bob Balling, of Arizona State University. The University of Virginia's Bob Davis, who wrote the initial text (since modified by the authors) for some of the sections on nor'easter storms and weather-related mortality, is another New Hoper. The same is true of for Arizona State University's Keith Idso, who wrote some of the original text on planetary greening, and Stanford University's Thomas Gale Moore, who contributed the material for a portion of the section on human health. For a much more in-depth look at that issue, consult Moore's fine book *Climate of Fear*, which Cato published in 1998.

The best New Hoper of them all is Paul C. ("Chip") Knappenberger, who did most of the technical analyses and graphics that went into this book. Chip is an encyclopedia of climate data and the best employee I have ever had. Amy Lemley, New Hope's editor extraordinaire (and maybe the best writer in Charlottesville), spruced up much of the text. Among other books, Amy is coauthor of *Beyond Shyness*, the title of which reflects the tone of this manuscript.

Considerable support for the quantitative analyses that went into this book came from the Greening Earth Society, of Arlington, Virginia. That membership organization was started by Frederick Palmer, CEO of Western Fuels Association, a not-for-profit energy cooperative. Without Fred, I truly believe the onerous Kyoto Protocol on global warming—or something like it—would be the law of this land today. The environmental community apparently agrees, too, judging from the number of television documentaries and newspaper features explaining that the only reason Americans have not cheerfully committed economic suicide to stop global warming is a pernicious industry campaign organized by Fred. Never mind that Washington spends at least *10,000 times* as much as industry to promote its view of climate change, a tribute to federal efficiency.

Greening Earth Society also supports the biweekly *World Climate Report*, the sassiest, hippest, nastiest, and best online newsletter in history on global climate. I edit the thing. It is available at www.greeningearthsociety.org. The *WCR* staff accomplishes at least two tasks—writing and communicating science—much better than the feds, and the newsletter therefore drives them to apoplexy and even worse writing. Perceptive readers will see that small portions of the text in this book look a lot like *WCR*.

Thanks also to the University of Virginia for allowing the release time for me to pursue this project. It is easy to gripe about political correctness and the leftward bias of the academy, about which there is little doubt, but it is also true that UVA has never impeded my freedom of expression and in fact has provided considerable internal funds toward getting my peculiar message out. It has not hurt that most of the climatologists at UVA are pretty much in agreement as to the nature of the global warming issue, and that Thomas Jefferson, whose presence still lives there (a friend said he saw him just the other day at the 7-Eleven!), fancied himself a climatologist too.

The cliché about thanks to my family for all the sacrifice while I put this manuscript together is more than true. Little can make one crabbier than writing a book, and many, many nights in Washington created this one, providing far too little attention to QTII, Bobbo, and Erika. You know who you are and why you are so special to me. It is my privilege to know that, too. Raoul the Siamese Weather Cat died for this book.

PJM
New Hope, Virginia
December 1999

Preface

In 1991 Patrick Michaels wrote *Sound and Fury: The Science and Politics of Global Warming,* and Robert Balling wrote *The Heated Debate*—both in response to the developing distortion of scientific reality by the political process and the media that cover it. Since then, that distortion has waxed and waned, always seeming to reach a crescendo prior to some important meeting, such as the December 1997 UN Conference in Kyoto, immediately before which ensued an onslaught of one-sided reporting. Because global warming and the Kyoto Protocol are likely to be important subjects in the 2000 election, now is an appropriate juncture at which to investigate the scientific developments that have emerged in the eight years between manuscripts.

Sound and Fury concluded with brief forecasts for the future—all of which, right or wrong, are encapsulated here. First, those that proved correct:

- **"The Temperature Record Will Warm....** We can, at least in the near term, expect to see news stories every January about how the preceding year was the warmest, or nearly the warmest, year in the past century."
- **"The Planet Will Continue to Warm....** There is little doubt that we have had some warming over the past 100 years.... The greenhouse enhancement is likely to contribute to warming."
- **"Illusion of Rapid Warming....** The combination of El Niño events, along with artificial urban warming and a slow real warming, will create the illusion of some brief periods of apparently rapid warming.... A good time ... might be in the mid-1990s, after Mt. Pinatubo's ash disappears and a strong El Niño roars in."
- **"Emissions Reduction Treaty....** The United States will agree to a carbon-emissions reduction treaty.... The cost will be several trillion dollars to a nation with a substantial debt."

- "We will see longer growing seasons, summer temperatures that do not change much (except that summers will tend to be longer), warmer nights, not much change in day temperatures, and a greener planet."

The following were forecast to occur on a time frame longer than the eight years between *Sound and Fury* and now:

- "The warming that will have occurred between 1900 and the time CO_2 effectively doubles in the next century will be on the order of 1.0°C to 1.5°C (1.8°F to 2.7°F)."
- "The costs of a carbon treaty will prove to be enormous, but the revenue generated will be addictive to government."
- "Historians in the 21st century will note that, even by the mid-1980s, the best available data indicated that the then Popular Vision of climate catastrophe was a failure."

So far, so good. Only the following *Sound and Fury* projection was wrong:

- Al Gore or George Mitchell would likely be nominated for president in 1996, running against Dan Quayle, Michaels predicted, anticipating that Gore or Mitchell would stand up in front of every drought or weather event and blame it on global warming, and that they would ridicule Quayle's stand on the subject. To be fair, *Sound and Fury* was written a few months after the Gulf War, when Bush's approval ratings were in excess of 90 percent; it seemed logical to assume he would be reelected in 1992.

That forecast was a near miss. Of course, Al Gore is running for president now, and for much of 1998 he was blaming every weather tragedy he could find—and the United States normally is replete with them—on global warming. It is an effective campaign strategy: Vote for me or you might die.

Overview

Global warming is a darned interesting scientific problem being played out as political drama, and this book is true to that model. We begin with the political theater, so common in recent years, in which the president or vice president insinuates that global warming is responsible for the weather tragedy du jour. In the United States, owing to our peculiar geography, we have the most violent combination of weather elements—snowstorms, hurricanes, tornadoes—in greater frequency than any other nation on earth. We are a nation of weather-watchers. Nowhere else in the world is there a profitable television network devoted exclusively to weather. Weather makes news, and news makes political opportunity.

Consequently, this book begins with the raging flood of the Red River of the North in 1997 and the attempts to relate it to global warming. Several other examples follow. In fact, many events that global warming is reputed to cause should become less frequent as the temperature rises.

Chapter 2 traces the development of global warming as an international issue, ending with the release of the second overall scientific assessment of the problem by the United Nations Intergovernmental Panel on Climate Change (IPCC) and the controversy surrounding that report. Many scholars and writers have dwelt upon the insertion of the statement that "the balance of evidence suggests a discernible human influence on global climate"—a phrase inserted after the full scientific review process—as the signal aspect of that report. We think that a much more obscure statement—that the climate models that initially fired up the concern about global warming were incorrect—is much more telling. This admission of exaggeration (or dare we say error) indicates that something may be very, very wrong with the glib projections of climate catastrophe that form the backdrop for the current political theater.

We cannot consider the merits and implications of the IPCC's statement about these models without a discussion of the nature of climate and climate change (chapter 3). We proceed much the same way most climatology courses do—although with mercifully less mathematics—from the consequences of the sun's differential heating of the earth to the major global circulation systems such as the jet stream and the trade winds. El Niño appears as a hybrid feature in which changes in the ocean, rather than the direct solar influence, create large-scale weather and climate consequences in some regions and overblown climate hype in others. Smaller-scale features such as hurricanes and tornadoes, resulting from the larger circulations, then enter into the discussion.

Having established the earth's climate from a descriptive point of view, we then turn, in chapter 4, to various models of climate change. These range from the empirical, which argues that the greenhouse effect has been altered for so long that nature has already given us the answer about global warming, to the hypothetical, in the form of megacomputer-driven "general circulation models" (GCMs). In their early incarnations, these latter analyses drove much of the initial concern and political posturing about global climate change. Indeed, these climate models started a process that ultimately resulted in a global climate protocol that the U.S. Senate is never likely to ratify.

Chapter 5 asks the deceptively simple question, "Has the earth warmed?" The answer surely depends upon when and where you look. In the broadest perspective, global warming is a very real thing, undeniable from surface temperature readings taken over much of the planet in the last 100 years. But a close inspection reveals a shift in warming patterns between the first and second halves of this century. Now, more than anything else, it is the most frigid air that is warming. The summer temperature rise is less than one-third of the change observed in winter. In the winter, the strongest warming is confined to the deadliest air masses, those over northwest North America and Siberia. And for the last quarter of this century, a peculiar anomaly has arisen: 80 percent of the lower atmosphere— the slice from just off the surface all the way to the stratosphere— has not warmed at all (except as a result of the now-departed mega –El Niño of 1998), as three independent measures of global temperature show.

Within the past few years, climate scientists have recognized that the original models were ill conceived and produced too much warming, too fast. To compensate, climate modelers have "tuned" their

GCMs with emissions other than carbon dioxide in an attempt to reduce the warming the models produce. Chapter 6 critically examines whether or not that effort has been successful and concludes that it has not. The alternative explanation—that the models simply overestimated the warming, rather than that something was hiding it—is increasingly attractive. Chapter 6 concludes with a forecast for global warming that results from an integration of the functional response of climate—as the models define it—with actual temperatures observed. Assuming a constant sun, we find that planetary surface warming should average around 1.3°C (2.3°F) in the next century, with the majority of that warming appearing in the winter. This number is close to the lowest end of the range the IPCC gave in its first report in 1990 and may allow for a common ground where perceived "skeptics" and "mainstream scientists" (we characterize ourselves as "mainstream skeptics") can all shake hands and go home.

Like chapter 3, which looks at climate circulation systems in relation to hurricanes, droughts, and flooding rains, chapter 7 compares model-projected changes in the climate circulation systems with what happens in reality. The results are largely consistent with those in chapter 6: The closer you look, the worse things appear. Virtually every climate projection forecasts increasing drought; yet the data show increasing wetness. Some models show increasing hurricane strength; yet maximum winds, measured for the last 50 years, are in fact declining along the U.S. East Coast.

Chapter 8 deals largely with the ocean, especially the notion of sea-level rise. As recently as 1980, the refereed scientific literature—the research papers that undergo peer review before publication—entertained projections of 10 feet or so—enough for the Potomac River to rise up and ring the base of the Washington Monument several blocks away. Since those early forecasts, projections have been scaled way, way back, with the IPCC's most recent summary providing two alternative scenarios of 10.6 and 19.6 inches in the next century—less than a foot to a foot and a half rather than 10 feet. But as chapter 6 shows, the warming projected to create these median estimates is probably still too large, so our corrected estimates come in even lower. Perhaps more important, however, is that natural processes have caused rises of similar magnitude to occur in the past century in some places, with little apparent disaster.

3

Because El Niño is largely an oceanic phenomenon, the chapter also critically examines the glib assertion that this recurring climate phenomenon is normally responsible for much of our nation's weather. That does not exempt the really huge El Niños, such as 1998's, which actually saved a lot of U.S. lives and money.

In chapter 9, we make the transition from the earth's climate to human life with a brief overview of the effects of climate change on mortality, with help from University of Virginia climatologist Robert Davis and Stanford University economist Thomas Gale Moore, who contributed a portion of the original text. Some facts: More people die from weather-related causes in the winter than in the summer. And per capita summer mortality is going down, thanks largely to air conditioning; from this perspective, proposals to fight global warming–related deaths in ways that make electricity more expensive appear inefficient, to say the least. (Moore's 1998 book, *Climate of Fear*, offers an extensive discussion of this subject, and we chose not to reinvent his elegant wheel here.)

As for plant life, chapter 10 draws on material from Arizona State University's Keith Idso to explore the stunning notion that we may actually be "greening" the earth via our increasing carbon dioxide emissions. Thousands of experiments in the refereed scientific literature show that plant growth is enhanced as that basic ingredient for photosynthesis rises in atmospheric concentration. We end with a look at global crop yields, which keep ascending despite decades of predictions to the contrary. The rate of yield increase seems remarkably insensitive to global temperature trends, although the big, warm El Niño year of 1998 is the world's record holder for agricultural production.

These 10 chapters form a large and internally consistent body of evidence against the prevailing view of climate gloom and doom. Each of us has summarized them into portable lectures that are as amusing to the local Rotary Club as they are to the American Meteorological Society—yet strangely disheartening to those who would have us believe that climate change portends the end of the world as we know it. Throughout the country, we have encountered a typical question at the end of each talk: "Gee, that made sense. So how come you are the only people saying it?" In truth, we are not. But it is correct to surmise that a larger number of people have a different and, we think, not so consistent point of view.

That said, chapter 11 attempts to answer this question by examining the dynamics of science and scientific support, marrying two powerful and highly cited ideas. The first is paradigm theory, espoused in Thomas Kuhn's *Structure of Scientific Revolutions*, which posits that most scientists work in support of a large conceptual framework, rather than working against one. The second is James Buchanan's Nobel Prize–winning work (with Gordon Tullock) on Public Choice theory, as articulated in *The Calculus of Consent*: When it comes to politics, he argues, we all behave in our own rational economic interest. Taken in combination, these notions suggest that when large amounts of monopoly funds are tied with political strings (the federal government is virtually the sole provider of climate research funding), they necessarily create a paradigm that temporarily distorts scientific reality. Shocking? No. Nor does it prevent good, "alternative" science from being published; rather, it has the perverse effect of making the "counterparadigm" research that does appear in the refereed journals less frequent but unusually strong.

Paradigms have consequences. Newtonian physics gave us ballistic missiles and then arms control treaties. Modern evolutionary theory yields treaties on biodiversity. The paradigm of rapid and damaging climate change gave us the United Nations Framework Convention on Climate Change in 1992 and the onerous Kyoto Protocol to that treaty in 1997; we detail this evolution in chapter 11, demonstrating that the Protocol will produce no detectable changes in global warming.

Chapter 12 presents our fearless forecast for the next 50 years. We admit we have no idea what society will look like in 100 years— could scientists of a century ago have anticipated space travel, the personal computer, or even a car in every garage?—so we do not take our predictions to the end of the century. "Global warming?" we conclude, "Get over it!" We close with a modest proposal for removing the melodrama from the theatrics guaranteed to surround the next great environmental threat.

1. The Shared Vision of Hell

Pieter Brueghel could not have painted a more lurid scene than what appeared on the evening news of April 13, 1997, as raging floodwaters rampaged through downtown Grand Forks, North Dakota. Entire blocks were aflame as rupturing gas lines collided with inevitable sparks. Not a fire truck was to be seen; none could ford the Red River of the North, normally a modest stream with little depth, now swollen to a mile wide and stories high.

Never had Grand Forks seen such a flood, and never would the city be the same. The vibrant town that served the fertile wheat-growing region of eastern North Dakota and western Minnesota became a shell of burned-out buildings and homes. Levels of unemployment and family misery grew to rival those of the Great Depression of the 1930s.

President Clinton blamed global warming. Prior to his April 22 flight to Grand Forks, Clinton spoke to reporters from the Rose Garden: "We do not know for sure that the warming of the earth is responsible for what seems to be a substantial increase in highly disruptive weather events, but I believe that it is."

Five months later, Vice President Gore dragged a quailing covey of reporters up to Grinnell Glacier in Glacier National Park. Speaking slowly, just to make sure everyone understood, he intoned, "This glacier is melting." As are many of the world's glaciers, he added, because of human-caused global warming.

A month later Gore was in California to lead the oxymoronic "El Niño Summit," where he was happy to conflate that natural oscillation in tropical Pacific sea-surface temperatures with global warming.

A few years earlier, in March 1995, Gore gave his annual Earth Day address at George Washington University. "Torrential rains have increased in the summer in agricultural regions," he said, referring to a yet-to-be published paper by federal climatologist Tom Karl. In fact, Karl had found no change in the frequency of daily

7

rainfall in excess of three inches. What he did find was a tiny change in the amount of rain coming from summer storms of between two and three inches in 24 hours, but these are hardly "torrential" and are most often welcomed by farmers everywhere, who pray for such rains. America's breadbasket is usually in great need of moisture come August.

In July 1998, Gore visited northeast Florida, which had experienced a series of substantial range and forest fires. He said the conflagrations "offer a glimpse of what global warming may mean for families." The reason Florida went up in smoke during this normally hot season was the overabundance of vegetation that resulted from excessive rains in the previous winter. While it might be convenient to finger the 1997–98 El Niño as the cause, statistical studies show El Niño is in fact associated with less-than-average burned acreage in Florida.

Gore's history of exaggeration—climatic and otherwise—is long and deep and repetitive. It begins long before his 1992 bestseller *Earth in the Balance: Ecology and the Human Spirit*, in which he says that fighting global warming is the "central organizing principle for civilization." He believes global warming is a battle between good and evil—which, judging from his past actions, includes anyone who disagrees with him about global warming. Referring to global warming in a 1989 article in the *New Republic*, Gore wrote, "'Evil' and 'good' are terms not frequently used by politicians. But I do not see how this problem can be solved without reference to spiritual values." This level of bombast and exaggeration (politicians use the words "good" and "evil" about as often as they say "children") has become a Gore pattern that now imperils his political future.

According to the Clinton-Gore administration, hot air even causes cold air. On February 5, 1996, many locations in the upper Midwest of the United States set their all-time records for low temperature (note that records in this region do not generally exceed 100 years in length). Two days later, while visiting the also-shivering schoolchildren of New Hampshire, President Clinton remarked that cold was the kind of thing caused by global warming. (Physically speaking, the administration here is apparently trying to repeal the first law of thermodynamics, which states that heat causes warmth and lack of heat causes cold.) Nine days later, speaking into a howling storm in Wilkes-Barre, Pennsylvania, Clinton blamed the snow on—what else?—global warming.

The list of these assertions is long. Taken together, they form a vision of hellish climatic catastrophe, a vision that has started to take hold. Sen. Larry Craig (R-Idaho) polled his constituents and found that 73 percent believe global warming is a real problem requiring real action.

Clinton's fallen political guru Dick Morris agrees. Appearing at one of Rep. Jack Kingston's (R-N.C.) regularly scheduled "Theme Team" meetings in fall 1998, Morris bragged that he had done some polling on global warming and said that if he were advising Gore in his presidential campaign, he would make that issue the centerpiece. In a departure from his normal mode of dispassionate analysis, Morris also stated that he truly believed global warming was a terrible problem. Morris thinks the American people share Gore's vision of climate hell, and that they believe it enough to elect him president.

How did this vision come about? And more important, is it true? *The Satanic Gases* holds the answer.

The science of global warming cannot be viewed outside the context of the "way science works," which Thomas Kuhn described in his 1962 classic, *The Structure of Scientific Revolutions*. Almost all scientists, Kuhn says, spend their lives doing "normal science," which includes the performance of simple experiments that verify that the current "paradigm" for a field is indeed correct.

Kuhn writes,

> Normal science, the activity in which most scientists inevitably spend almost all their time, is predicated on the assumption that the scientific community knows what the world is like. Much of the success of the enterprise derives from the community's willingness to defend that assumption, if necessary at considerable cost. Normal science, for example, often suppresses fundamental novelties because they are necessarily subversive of its basic commitments (p. 5).

In Kuhn's world, scientists toil under overarching structures, or "paradigms," and "normal science" consists of shoring up any little problems or inconsistencies within that structure:

> Mopping-up operations are what engage most scientists throughout their careers. . . . Closely examined, whether historically or in the contemporary laboratory, that enterprise seems an attempt to force nature into the preformed and relatively inflexible box that the paradigm supplies (p. 24).

9

A "paradigm" is, for example, the earth-centered universe, defended in its day by academic scholars everywhere. When Galileo looked at the moons of Jupiter, it could have been but a few minutes before he realized they were more analogous to the relationship of the earth to the sun than the existing paradigm. But that irritated most official (i.e., church sponsored) scientists at the time, so Galileo found himself being hauled before the Inquisition, which threatened him with death. When Gore was in the Senate, he merely hauled paradigm-smashers before his "Science Roundtables" and threatened them with discredit.

Kuhn notes that when a paradigm is threatened by inconvenient data, the first response is to ignore reality:

> In science . . . novelty emerges only with difficulty, manifested by resistance, against a background provided by expectation. Initially, only the anticipated and usual are experienced even under circumstances where the anomaly is later to be observed. Further acquaintance, however, does result in an awareness of something wrong or does relate the effect to something that has gone wrong before (p. 64).

"Normal science" in the greenhouse issue is the notion that computerized climate models are producing a largely realistic picture of the atmosphere warmed by carbon dioxide, if twiddled a bit here and there ("mopped up," in Kuhn's view), and that this warming will be rapid and disastrous. In the longest run, though, Kuhn is predicting that something will eventually be found to be gravely wrong with the current paradigm—a proposition we detail in succeeding chapters.

Scientists also function within a larger society and respond not only to their Kuhnian dictates. They also have personal, ethical, and financial interests, just like everyone else. James Buchanan has described the interaction of these interests in the public sphere under the rubric of "Public Choice Theory," and the combination of Buchanan's theory and Kuhn's theory goes a long way toward explaining the history of global warming science, a discussion we save for chapter 11. It is first necessary to talk about global warming itself and the scientific basis for projections of future change.

2. Global Warming Goes Global

The idea that human beings could change global climate developed in the 19th century, with the realization that certain industrial emissions—notably carbon dioxide—could alter the rate at which heat escaped from the lower atmosphere. But, in fact, people have changed climate ever since the first hominid cleared a patch of land.

Only the scale has increased. As agriculture radiated away from the cradle of civilization, human beings began to alter the surface of large regions on the planet. Entire ecosystems have been removed, such as the perennial long grass prairie of central North America, which has been replaced with annual corn and soybeans. Whereas the prairie was a continuous vegetative cover, the replacement crops are seasonal, with bare ground exposed to the sun for half the year, resulting in dramatically different absorption of and heating by the sun's radiant energy.

By the dawn of the 20th century, regional alterations, such as the burgeoning economy of the United States, began to exert global influence on the climate, either because regional temperature changes caused by land surface alterations must eventually spread into the overall climate system, or because the emissions resulting from economic activity—among them carbon dioxide—spread throughout the planet.

Humans' dynamic changes to land form are now thought to be perhaps as important as the alterations of the greenhouse effect in determining climate change, according to Colorado State University meteorologist Roger Pielke and vegetation modeler Gordon Bonan. But amazingly, land use changes are scarcely considered by the computer models that serve as the basis for the current policies. Instead, carbon dioxide dominates the discussion.

Thus the dominant paradigm in climate change has somehow largely ignored one of the most important determinants of human influence on the atmosphere. At the same time, self-selected communities, usually under some governmental aegis, began to attempt

11

explicit definitions of the global warming paradigm. In 1986, the U.S. Department of Energy released a five-volume "state of the art" (dated 1985, but published a year later) compendium called *Carbon Dioxide and Climate Change*.

There have been many such documents. Those produced subsequently, mostly by the United Nations, have repeatedly demonstrated that very little has changed scientifically in the last 15 years. A summary of the 1985 DOE report (and the many United Nations follow-ups) might run as follows:

1. *The earth's climate is complicated.* Though it is easy to calculate the gross temperature of the planet as a whole, we have only limited understanding of what defines the climate at any given location. As a result, our knowledge about how climate changes over time at any location is even more inadequate. The only way we know how to improve our understanding is with the use of increasingly complex computer models that attempt to simulate the atmosphere's behavior over long time and space scales. The belief that this is indeed even possible and that the current models are realistic is the reigning paradigm.

2. *The earth's surface temperature is not constant.* For reasons that are not at all clear, the mean surface temperature of the planet varies by about 5°C (9°F) on the 100,000-year time horizon and has been doing so for millions of years. On the 1,000-year scale, variation is about 1°C to 2° C (1.8°F to 3.6°F). In the last 10,000 years there were two well-known excursions of temperature. One was a warming on that order that occurred around 5,000 years ago, and another was a cooling of about 1°C (1.8°F) that occurred in the past 1,000 years and from which we have only recently emerged.

3. *The earth's surface temperature has warmed in the last 100 years.* Although there are problems associated with measurement, such as the well-known fact that heat-retaining concrete cities tend to grow up around their weather stations, supplying an artificial "urban warming," the mean surface temperature has warmed a bit more than 0.6°C (1.1°F) in the last century.

4. *Human activity is changing the composition of the atmosphere in ways that may affect climate.* Certain atmospheric constituents, notably carbon dioxide and methane, are increasing. Everything else being equal, these compounds will warm the atmosphere below around

45,000 feet and cool it above that level. (The next chapter deals more extensively with this process, commonly called the greenhouse effect.)

5. *Observed temperatures are "not inconsistent" with computer models that attempt to simulate the changed greenhouse effect.* This artful choice of words in the 1985 Department of Energy report means that the earth's temperature is going up, not down, and nothing more. If it were going down, the inconsistency would be obvious. But the rise is merely "consistent" with greenhouse changes. It is also "consistent" with the global cooling trend of the last 1,000 years that has reversed only in this century, as climatologist Michael Mann recently showed in *Nature* magazine.

These DOE reports started a process through which science evolved into public policy on global warming.

The 1985 IIASA Conference

The first attempt at internationalizing global warming science and policy took place in 1985, when the International Institute for Applied Systems Analysis, in Laxenburg, Austria, sponsored a joint meeting of the UN Environment Program (UNEP) and the International Council of Scientific Unions on global warming. IIASA had long been suspected of being a convenient interface between the then-communist world and the West, and many of the contacts made at that meeting remain productive today. The David Suzuki Foundation, a radical group promoting green causes in Canada, considers IIASA one of the "world's key sustainable development organizations." A visit to its homepage (www.iiasa.ac.at) confirms IIASA's agenda.

Together, the 1985 IIASA attendees concluded that "the understanding of the greenhouse question is sufficiently developed so that scientists and policymakers should begin an *active collaboration* [emphasis added] to explore the effectiveness of alternative policies and adjustments," and to "initiate, if deemed necessary, consideration of a global convention." Thus the notion of a climate treaty, driven by a handful of concerned scientists, was born in Laxenburg. As Professor Frits Bottcher of the Global Institute for the Study of Natural Resources in the Netherlands noted, "The scientists who attended the meeting had insufficient evidence for the statements"—an understatement that will become obvious in succeeding chapters.

Daniel Bodansky of the University of Washington wrote that it is highly questionable that the state of the science at the time of the IIASA conference could have propelled the issue as rapidly as it actually moved. Instead, he wrote, "a number of scientists . . . acted as entrepreneurs, promoting the climate change issue."

UNEP was founded at the height of the last global "ecology" enthusiasm, in 1972, at the first United Nations Conference on the Environment in Stockholm. As one of the principals in the 1985 IIASA conference, UNEP gave the United Nations entrée into the global climate issue. What could better suit the United Nations' consistent agenda of wealth transfer? Here, finally, was a truly global issue in which the alleged perpetrators—the rich, industrial North—could be blamed for the coming environmental Armageddon. Given the voting structure of the UN General Assembly, as long as the United Nations could generate some type of scientific cover, global warming would muster a majority sentiment for transferring money from North to South.

Even better, residents of the North who were generally supportive of the United Nations' internationalist agenda also were highly attuned to the global warming issue. Foremost among these was rising star Al Gore, a preeminent believer in the importance of international global oversight of environmental issues. Gore progressed rapidly from representative to senator to vice president and then presidential candidate—along the way serving as the Senate godfather for massive amounts of federal science funding in support of the UN view that climate change was a serious problem.

The Intergovernmental Panel on Climate Change

By 1988, UNEP and another UN entity, the World Meteorological Organization, had combined to form the UN Intergovernmental Panel on Climate Change (IPCC), which described itself as "an intergovernmental mechanism aimed at providing the basis for the development of a realistic and effective internationally accepted strategy for addressing climate change."

This description speaks volumes about the IPCC's purpose. It describes itself as a governmental "mechanism" to provide the "basis for...a strategy to address climate change." In other words, to its members, disastrous climate change (which most people translate as "global warming") is a given, as is the need to do something about it, unless of course you believe that people in government think they are there to do nothing.

The IPCC was more than merely predicated upon the belief in human-induced climate change. It was in fact directed to produce a basis for a treaty, which became the Framework Convention on Climate Change (FCCC) described in chapter 11. In 1988, the UN General Assembly voted that the IPCC was "to initiate action leading as soon as possible to recommendations with respect to the identification and possible strengthening of relevant existing international legal instruments having a bearing on climate [and] elements for inclusion in a possible future international convention on climate."

Which is to say, if the IPCC were to state that climate change was not a dire threat requiring a new United Nations treaty, it would be disobeying the very orders under which it was created.

The panel has produced three summary volumes on the issue of climate change, in 1990, 1992, and 1996. Given its charter, who can honestly believe that the IPCC's self-selected participants would state that climate change might not be a problem? Why would an organization choose members who were inimical to its statement of purpose?

So the IPCC was designed to support a foregone conclusion— something that its proponents, such as Gore, seem to ignore when they cite it with the well-worn aphorism "the consensus of the world's best climate scientists." In fact, it largely represents the consensus of scientists who attended the 1985 IIASA meeting, contributed to the 1986 Department of Energy "state of the art" volume, and were selected by their governments to work for the United Nations.

Then there is the entire army of IPCC participants who are known only as government functionaries. For example, the first (1990) report, *Climate Change: The IPCC Scientific Assessment*, lists three main

15

authors. Only one, John Houghton, who was the IPCC's chief scientist, is a climate physicist.

IPCC Leader Mixes Science and Religion

One thing scientists try to do is to keep the supernatural out of the natural world they so objectively study. But it is clear that John Houghton, senior scientist for the IPCC, from the organization's inception to the creation of the onerous climate treaties, has had no problem mixing the two.

On June 17, 1996, as a run-up to an important UN Geneva meeting whose goal was to strengthen the climate treaty that subsequently grew out of the IPCC (chapter 11), Houghton wrote in the *London Times* that climate change is a "moral issue" and that he welcomed "the current initiative of the World Council of Churches which calls upon the Government to adopt firm, clear policies and targets, and the public to accept the necessary consequences." This sentiment echoes Al Gore's words in *Earth in the Balance* about "the public's desire to believe that sacrifice, struggle, and a wrenching transformation of society will not be necessary" to fight global warming.

Houghton went on to say that reducing greenhouse gas emissions will "contribute powerfully to the material salvation of the planet from mankind's greed and indifference." Here he conveniently wedded his scientific authority, as an influential member of the IPCC, and religion, to create some virtuous privation and "material salvation."

In devising the IPCC roster, members of the UNEP submitted lists of scientists to the World Meteorological Organization (another UN body) secretariat and a group of senior administrators including John Houghton selected the final list. One thing is certain: there are very few degreed climatologists in the IPCC. Prior to its founding in 1987, very few doctorates were awarded in climatology because the area was viewed as the lowest-prestige specialty in atmospheric science. Now, of course, almost all atmospheric scientists want to find a way to call themselves climatologists, because that is where the funding (and therefore the prestige) resides.

The list of authors and reviewers of the IPCC documents includes just a small number of individuals specifically trained in climatology. In the 1990 document, 70 of the 214 listed (33 percent) "lead authors" and "contributors" appear to meet this qualification.*

The environmental community is fond of labeling us and our friends a "small band of scientific skeptics" (usually numbering around 10). But the scientific "mainstream"—the 70 or so other bona fide climatologists in the IPCC—is at best an only slightly larger "band." A more accurate description of us would be that we are "mainstream skeptics," whereas many of our remaining colleagues are the IIASA scientists Daniel Bodansky referred to who "acted as entrepreneurs, promoting the climate change issue."

The IPCC's *Scientific Assessment* was nothing really new inasmuch as scientists have been summarizing their views on this and other scientific issues for decades. For example, the *World Survey of Climatology*, a 14-volume behemoth published over several years in the early 1970s, is much more comprehensive, although it seems more evenly disposed to global warming, global cooling, an ice age, or more of the same.

The key sentence in the entire 302-page IPCC document is this one: "When the latest atmospheric models are run with the current concentrations of greenhouse gases, their simulation of present climate is generally realistic on large scales."

That sentence defines the paradigm that lit the global warming fuse. In this case, the "latest" models refer to the large computer climate simulations, called General Circulation Models (GCMs), in which the natural greenhouse effect was altered in a fashion thought to mimic changes induced by human activity. (Chapters 4 through 8 detail the evolution and reality of these projections.)

The 1990 report predicted that the global average surface air temperature should be rising at a rate of 0.3°C (0.5°F) per decade. It

---

*That the majority of "players" in the climate change issue were not atmospheric scientists became apparent to Michaels at a 1992 meeting in Germantown, Maryland, sponsored by the U.S. Department of Energy. The gathering's purpose was to figure out how to spend tens of millions of taxpayer dollars researching global climate change. At one lunch, he asked for a show of hands at his table of 12 attendees from all who held academic degrees in meteorology or climatology. Two were raised. Yet the IPCC had designated everyone at the table as a climate expert.

further predicted that warming would be greater over land than over the ocean and greater in the winter than in the summer.

The IPCC issued a special supplement in 1992 specifically to provide the technical background for the United Nations' Framework Convention on Climate Change, to be proposed at the Rio "Earth Summit" beginning June 1 of that year. According to Bo Kjellen of the Swedish Ministry of the Environment, the Framework Convention existed because "the Intergovernmental Panel on Climate Change had presented unequivocally the case for early action."

In fact, the 1992 supplementary report did more. For the first time, the IPCC was lowering its estimates of future global warming because of the well-known (to scientists) but previously concealed (from policymakers) discrepancies between computer forecasts of climate change and observed temperatures.

But the process went forward at its own peril. Six years later, in July 1998, the new head of the IPCC, Robert Watson, would admit in a congressional hearing that the models that were the basis for the 1990 report were, in his word, "wrong." How could the IPCC invalidate its own "consensus" in such a short time?

The 1996 IPCC "Second Assessment"

The IPCC's original 1990 *Scientific Assessment* of global warming and the subsequent 1992 *Supplementary Report* provided scientific covering fire for the Framework Convention. A new report was needed by the mid-1990s, as the July 1996 meeting of the signatories in Geneva was scripted to produce an apparent consensus for "legally binding" commitments to reduce greenhouse gas emissions on the part of a few wealthy nations. A subsequent meeting, scheduled for Kyoto in 1997, was supposed to devise actual language that would amend the Framework Convention to become "legally binding."

To get maximum press, the IPCC placed the draft version of its not-yet-finally-reviewed Second Assessment Report on the Internet in late summer 1995. By Sunday, September 10, it was the headline above the fold in the *New York Times*.

The article had "Draft—Do Not Quote or Cite" plastered all over it, which is a green light encouraging any good reporter to do just the opposite, and Bill Stevens of the *Times* obliged. He said at the time, "It was on the Internet, and therefore I felt I had to do it."

Stevens wrote of "an important shift of scientific judgment" and said the new IPCC report was much more emphatic than previous ones that humans were changing global climate.

The full report was published in mid-1996. There really was not much new in it; science cannot be expected to change much between the summer of 1992, the time of the previous report, and mid-1995. The "important shift" did not occur until two months after Stevens wrote about it, because at 20 minutes past midnight on November 30, 1995, long after the scientific peer-review process was finished, a meeting of a very small fraction of the IPCC approved a text change. They inserted this statement: "*The balance of evidence suggests a discernible human influence on global climate.*" The statement is actually devoid of meaning, because the important factor is *how* and *how much* humans change the climate, not whether they do.

According to climatologist Bruce Callendar, who was one of the report's senior editors, "All knew that the sentence would be the most widely quoted in the report." The statement was driven by a key paper that had yet to be published in the refereed literature, written by Benjamin Santer of the Lawrence Livermore National Laboratory. (The paper ultimately was published to severe criticism, as chapter 6 details.)

Callander then chose to cover the IPCC's extremities. He wrote, "Some of the media conveyed this message to the public in a corrupted form, presenting it as proof that greenhouse gas emissions . . . could be blamed for contemporary weather extremes, from floods to droughts."

He neglected to point out that the "corrupters" could well include the president and vice president of the United States—recall their performance when confronting floods and other weather extremes.

Does anyone seriously believe that Callander, who was savvy enough to know the "discernible human influence" statement would be the most widely quoted, did not also know exactly how it would be "corrupted," by the media and others? It is hard to imagine being a senior editor of a document that claims we can calculate the way the wind 100 years from now will blow through our impressively complicated atmosphere and not knowing the downwind trajectory of the "human influence" statement.

Nonetheless, a much more important observation in the 500-page report got "missed" by the eager press:

> When increases in greenhouse gases only are taken into account ... most GCMs produce a greater mean warming than has been observed to date, unless a lower climate sensitivity is used.... There is growing evidence that increases in sulfate aerosols are partially counteracting the [warming] due to increases in greenhouse gases.

This remarkable paragraph states the following:

1) The models used in the original IPCC assessment were wrong.
2) Unless something is "hiding" the warming, the sensitivity of the surface temperature to human greenhouse effect changes was overestimated—which is to say, it simply will not warm up as much as previously forecast. (That "something" was supposed to be a pollutant called sulfate aerosol, discussed extensively in the next chapters.)

Charges went back and forth about the insertion of the critical sentence about the "discernible human influence." Critics argued that the IPCC document was not really peer reviewed in the traditional sense of the term, since the authors had the ultimate right to discard or disregard any criticism if they so chose. The former head of the U.S. National Academy of Sciences (and president of the American Physical Society), Frederick Seitz, wrote in the *Wall Street Journal*, "I have never witnessed a more disturbing corruption of the peer-review process than the events that led to this IPCC report." The IPCC's John Houghton called Seitz's comments "just rubbish," and Santer darkly hinted about a conspiracy, saying, "One doesn't require much imagination to identify the underlying culprits," as if Seitz had committed some crime by expressing his considered opinion. Sounding paranoid, Santer said in *Science* that "I [Santer] am being 'taken out' as a scientist." "Powerful interests," he claimed, "were intent on skewing the 'balance' [of the chapter he wrote], and on accentuating the uncertainties, rather than what we had learned in the past five years." This from the author who supervised the insertion of the most quoted line in the entire report, put in *after* normal peer review.

20

The new report even got its greenhouse gases wrong. When the IPCC made projections for increases in their 1996 comprehensive report, according to David Schimel of the U.S. National Center for Atmospheric Research (NCAR), who wrote part of it, they used projected increases in methane, an important greenhouse gas, that "were based upon an understanding that was five to 15 years old."

In all, the 1996 report has its good and its bad points. Santer may complain that some readers excerpt only one "side" of its interpretations, but that was probably what many of the authors intended. In fact, portions of *The Satanic Gases* will rely heavily on it. After all, the 11th-hour insertion of the "human influence" statement did not invalidate everything else contained therein. For the record, we too believe that there is a human influence on the climate. But, to put it simply, the effect is just not all that bad.

3. The Earth's Climate

The notion of a "changed" climate is predicated upon an understanding of the causes of climate and weather. While we are often pelted with the notion that the "climate problem" is the most complicated issue on the planet, actually it is not. The sun shines upon the earth, and the earth's components, including its oceans and its atmosphere, interact with solar radiation. That interaction is called "climate."

The Energy Spectrum
The sun warms the surface of the earth, and the earth dissipates that heat out to space. The sun's light is a form of radiation. So is the earth shining out to space. Our space probes, looking back, show us that, for color and brilliance, our planet is the glamour queen of the solar system. But to the dispassionate sensors on the satellite, it is just radiation.

Everything that is not at absolute zero (and nothing is) radiates energy. It is a law of physics that cold things radiate less energetically than warm ones. Sometimes it is convenient to display the radiation of the universe on a spectrum—from the lowest-frequency signals to high-energy thermonuclear signals like gamma rays (see Figure 3.1). Their power can be expressed as the wavelength of the radiation, with low-energy sources being of longer wavelength than high-energy ones. Many low-energy sources produce very familiar and useful radiation. Radio waves, for example, come from relatively low-temperature sources; they are so low-energy that they can be emitted on purpose—from radio transmitters—with virtually no discernible effect on living things.

Quantum physics yields something called the Wien displacement law, which states that the maximum wavelength at which a body radiates is a function of its average temperature. Warmer bodies radiate at shorter, more energetic wavelengths, and colder bodies at longer ones.

Figure 3.1
THE ELECTROMAGNETIC SPECTRUM

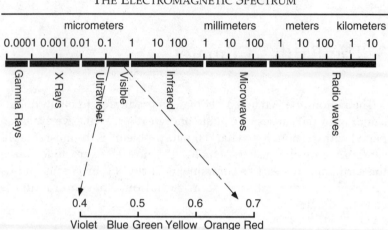

NOTE: The sun emits primarily in the visible and ultraviolet range, while the earth radiates primarily in the visible and infrared. Greenhouse gases trap radiation in the infrared range.

The earth's mean surface temperature of about 15°C (59°F) causes it to radiate primarily in the infrared and visible wavelengths, which are at the medium-energy level. Many gaseous molecules have a geometry that resonates with infrared wavelengths, causing them to absorb infrared radiation. This property turns out to be very important in any discussion of human-caused global warming.

Visible light, which clearly creates a lot more good than harm, is what plants harness for photosynthesis. Light drives forward a series of chemical reactions that take carbon dioxide and water out of the atmosphere and synthesize them into complicated organic molecules organized into plants.

Still farther up the spectrum is ultraviolet radiation—just a tad more energetic than visible light, but strong enough to slowly damage (burn) unprotected living surfaces. Most animals—humans being a notable exception—are largely covered with enough hair to shield the skin, and most plants have thickened surfaces that accomplish much the same.

The sun's mean surface temperature of about 6,000°C (10,800°F) causes it to radiate primarily in the visible and ultraviolet range. It should not be surprising that plants settled on a substantial portion of this visible range as the driving force for life; after all, it is the most abundant stream of radiation hitting the earth, according to the Wein displacement law.

Interestingly, the heat of the sun is not uniformly distributed around the planet or within the atmosphere. Obviously the polar regions are cold and the tropics warm, because the sun's rays are much more direct over the low latitudes than they are over Antarctica. The surface temperature is further complicated by the distribution of land and ocean (land heats more rapidly than water) and the color of the surface. White snow reflects away the sun's radiation, resulting in a cooler surface, while black (the effective color of the deep blue sea) absorbs radiation. The relative amount of radiation reflected away is known as the "albedo," and it is clearly something that humans can change, simply by paving a desert or a forest, or changing clothes from white to black.

The Greenhouse Effect

The heat of the earth also does not go directly out to space. Again, several different compounds in the atmosphere have a molecular structure that resonates at the same frequency with which the earth radiates the sun's energy outward. These molecules, including water vapor, carbon dioxide, methane, and several very minor constituents (such as manmade chlorofluorocarbon [CFC] refrigerants) intercept a portion of this energy and then reradiate it either up (out to space) or down (back toward the surface). The "downwelling" radiation is what is commonly known as the "greenhouse effect," and it makes the surface of the planet about 33°C (59°F) warmer than it would be without it.

Almost all of the natural greenhouse effect, about 31°C of the 33°C worth, comes from water vapor. The concentration of atmospheric water vapor varies with temperature, and its range is quite spectacular. Everything else being equal, the amount of water in an atmosphere of −40°C (−40°F) is about 1,000 times less than what resides at a temperature of +40 degrees C. That 80°C (144°F) range is about what we see on earth—from the cold of Antarctica to the Sahara Desert.

As long as the surface temperature of the globe remains constant, the amount of water vapor in the atmosphere stays the same. That is because water's evaporative pressure at the interface between the atmosphere and the surface (which is largely ocean) is a direct and quantifiable function of the temperature. For the purposes of this discussion, all of the water vapor in the atmosphere ultimately comes from the ocean (in fact, a tiny, tiny amount comes from the very occasional comet that breaks up aloft).

Most of the rest of the natural greenhouse effect comes from carbon dioxide, the atmospheric level of which can be changed by geological events such as volcanoes, biological events such as photosynthesis, and human events such as the combustion of fossil fuel. Conservative political activists and some industrial organizations often make the incorrect argument that carbon dioxide is unimportant because it is a relatively small contributor to the overall greenhouse effect. But the other large contributor is, for all intents, constant as long as the temperature does not change. And if an additional increment of carbon dioxide warms the planet, the increased evaporation from the ocean will admit more water vapor to the atmosphere, which theoretically also will raise the temperature.

The same activists often also contend that only a small fraction (about 5 percent) of the total amount of carbon molecules that goes into the atmosphere each year is from human activity, and therefore the human contribution is inconsequential. That is simply not true.

Nonhuman events that change the carbon dioxide greenhouse effect include not only volcanoes but also forest and prairie fires (fires temporarily reduce the amount of photosynthesis while also releasing carbon dioxide via the combustion process). But by and large, these are pretty random variables. No one has yet come up with a scheme that reliably predicts the number, location, and timing of big volcanoes decades in advance, and the burning of prairies and forests, though inevitable, is hardly constant. After, say, a big volcano, plants eventually take in, or "capture," a bit more carbon dioxide, animals eat the plants, the animals die, and the CO_2 is once again interred.

The human contribution of carbon dioxide has indeed been increasing for about 200 years, with the largest changes, about 75 percent of the total greenhouse effect change, occurring since World War II. Eventually the plants, animals, and ocean *will* capture and sequester that amount, but it takes time. Studies of atomic bomb tests show it takes about 10 years for vegetation to capture the

Figure 3.2
TIME HISTORY OF GREENHOUSE ENHANCEMENT

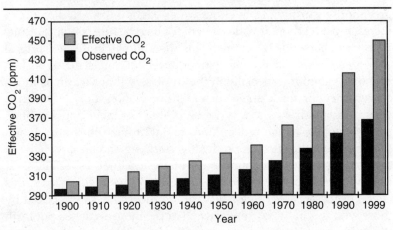

NOTE: Time history of the greenhouse enhancement. Totals are in effective concentration of carbon dioxide and reflect the cumulative effects of emissions of carbon dioxide, methane, chlorofluorocarbons, and other, more minor, greenhouse gases.

"average" molecule. But then some of that same carbon is recycled as the vegetation burns or is consumed and respired by an herbivore. No one really knows how long the average residence time turns out to be, but median estimates range from about 30 to 80 years. Environmentalists like to cite the latter figure, conveniently rounding up to a hundred years (so that the IPCC can call carbon dioxide a "century-scale" emission, as it did in its 1999 *Special Report on Aviation and the Atmosphere*). But the true median figure is considerably below this value.

At any rate, a simple look at the greenhouse effect and human activity says this: Put carbon dioxide in the atmosphere and it will warm up. Needless to say, things are not so simple.

One of the major complications has to do with water vapor and carbon dioxide's absorbing and reradiating much of the same wavelengths of infrared radiation emanating from the surface. There's only so much to go around, and an increasing density of either molecule results in successively less greenhouse warming being contributed on a molecule-for-molecule basis.

This is called the "water vapor–carbon dioxide overlap" and it has important consequences. Dry air will have very little natural greenhouse warming from either water vapor or carbon dioxide, so increasing its carbon dioxide content (or its moisture content) results in a very sharp initial heating. This then allows more moisture to evaporate, further raising the temperature. Very moist air, as is often found in summer, especially in the tropics, will not warm as much as very dry air for a given increment of carbon dioxide. Further, warm, moist air has a way of becoming unstable and bubbling up into clouds, which reflect more radiation than they absorb and therefore cool the surface. So the big warming should largely be in the coldest air masses, found most often in winter, especially over northwestern North America and Siberia. As we will see later, this is one of the aspects of global warming most underappreciated by those who sound the current alarm. For the moment, let us put forth the proposition that warming up Siberia in the dead of winter is probably a good thing.

Greenhouse Warming = Stratospheric Cooling

The greenhouse effect does not change the temperature of the entire earth-atmosphere system; it merely distributes the heat within that system. For the overall temperature to change significantly, our sole source of heat—the sun—must change. There is some evidence that this is the case, but greenhouse gases alone will not alter the overall energy balance.

The greenhouse gases are concentrated in the lower atmosphere—mainly in the earth's active weather zone known as the troposphere. The top of this zone averages around 50,000 feet in midlatitude summer and down to around 35,000 feet in midlatitude winter. The "greenhouse effect" is merely the capturing of some of the infrared radiation emanating from the sun-warmed earth before that radiation passes directly out to space. When greenhouse gases such as water vapor or carbon dioxide reemit this radiation, it is "recycled" in the troposphere.

So the lower layers of the atmosphere should warm with increased carbon dioxide, while the stratosphere, which sits over the troposphere, will cool because of the slightly decreased flux of radiation that has been recycled beneath it. Though some problems exist concerning the amount and distribution of measured warming in the troposphere, there is little doubt that the stratosphere is cooling.

Why the Earth Is Not Frozen

The trendy paradigm of "fragile earth" may dominate public discussion, but nothing could be further from the truth in terms of climate. That the planet is not in a permanent ice age is evidence enough.

Seventy percent of the surface of the earth is liquid water. And an additional 5 percent or so of the surface is solid water, mainly Antarctica, Greenland, and the icepacks of the polar and far southern oceans. Almost all the rest of the surface is vegetated. Subtropical deserts like the Sahara stick out as barren anomalies in satellite photos of the planet. Plants themselves are 70 percent to 90 percent water.

When sufficiently deep, liquid water on the surface of the planet is virtually black, meaning that it reflects away very little incoming solar radiation, especially when the sun is fairly high in the sky. The tropical oceans are especially efficient at absorbing this energy.

Ice is much brighter, reflecting away about 60 percent of the sun's heating rays, and snow does even better, reflecting 90 percent. So, once a large amount of water freezes, the earth's temperature should go into a "negative feedback loop," with increasing amounts of ice reflecting increasing amounts of radiation—and on and on until much of the planet looks like an iceball.

The current mean temperature of the planet—59°F (15°C)—is only 27°F above the freezing temperature of water, which makes up virtually all of its surface. Temperatures only have to drop a little, say, around 3°C (4.5°F), and the amount of time that snow lies on the ground at high latitudes increases dramatically. Then as the increasing whiteness takes over, the earth reflects away more and more radiation and temperatures continue to drop, resulting in an ice age. Temperatures cannot make it down far enough in the tropics to completely freeze things over, but much of the mid- and high-latitude continents can easily be covered with thousands of feet of ice.

(continued next page)

(continued)

At first, you might think the earth would easily run away into a frozen iceball, given these conditions. After all, the planet becomes increasingly white and cold. But it does not. We know this because we are here. If it had ever frozen over it would likely never have thawed, and evolution would have proceeded in a dramatically different fashion. That this never happened is quite amazing, given that the mean surface temperature is so very close to freezing. (A recent article, published in 2000 by Hoffman and Schrug, hypothesizes that the earth did freeze 600 million years ago, but the theory enjoys virtually no support from climatologists.)

While the last 2.5 million years or so have been punctuated by ice ages, these are rare in the geological sense. The vast majority of the last billion years were devoid of large ice sheets. Even with our current propensity for glacial ice, the climate is hardly "fragile," as it somehow always snaps back from the cold, white abyss of glaciation. Thus, the earth enjoys a remarkable stability, in spite of the great changes to the planet's whiteness that accompany ice ages. The massive differences in absorbed solar energy on land areas between a glaciated and a nonglaciated earth dwarf anything that human beings could directly induce by changing the greenhouse effect.

The Changed Greenhouse Effect

One thing concerning global warming about which there is no debate is the notion that human activity has augmented the earth's natural greenhouse effect. The magnitude of this change, coupled with a deficit of predicted warming, is what fuels the core of the argument that global warming is an overblown issue.

Between the end of the last ice age, about 11,000 years ago, and the dawn of the industrial revolution, roughly 200 years ago, the concentration of carbon dioxide in the earth's atmosphere has varied from around 260 ppm to about 320 ppm. In other words, in a million cups of air, there was the equivalent of an average of around 280 cups of carbon dioxide.

Returning to the subject of scientists' devotion to the prevailing paradigm, it is interesting to note that CO_2 concentration was thought to be quite stable, hovering around the 280 ppm mean. The notion of equilibrium was scientific gospel from 1958, when CO_2 concentrations were first measured in a rigorous fashion, until 1999. In that year, Friederike Wagner et al. published research in *Science* revealing that their detailed analyses of ice cores show a 60 ppm range. Wagner's finding means that plants—which remove CO_2—and the flux of CO_2 into the atmosphere from decay of living organisms have never really been in equilibrium. Obviously, the postglacial constancy of CO_2 before the industrial revolution is central to the global warming issue, and here an invalid concept was defended to the death for 41 years—complete with the branding of critics as ignorant louts—but it turned out to be simply untrue. Unfortunately, this is not an unusual occurrence in the history of science. One era's paradigm is often a future era's laughingstock, but the holders of the current paradigm cannot seem to recognize that fairly high probability.

Carbon dioxide concentrations have risen from the preindustrial range to a current value of about 365 ppm, or an increase of 30 percent. That figure is about 50 percent of the total human greenhouse change because other greenhouse gases have increased too.

The recent behavior of another greenhouse gas, methane (CH_4), a.k.a. "natural gas," has shattered another long-held myth about climate change. In its total absorption of infrared radiation, methane is the second most important of the anthropogenerated emissions. It currently contributes around 20 percent of the total human greenhouse change. Its current concentration in the atmosphere is about 1,700 parts per billion (ppb), and for much of the 20th century it has been rising at the rate of about 12 ppb per year, a substantial climb.

Wetlands are a large natural source of methane, so their putative disappearance would decrease methane concentrations. Coal mining releases the gas to the atmosphere—not surprisingly, since coal seams are close to fossilized wetlands. Bovine flatulence, a term the University of Virginia's Fred Singer coined more than 20 years ago, supposedly contributes a significant fraction, given that the number of bovines increases proportionally to human longevity. It is a fact that the longer people live, the richer they tend to be, and they like to eat steak. Longer lives = more steak = more bovine flatulence.

Another source of methane is thought to be leakage from natural gas pipelines. Communist-era pipes were thought to be especially leaky.

With the exception of wetland losses, all of these sources of methane suggest that its level in the atmosphere should be increasing in an exponential fashion (with the concentration rising at ever-increasing rates). That the increase had instead been linear (rising at the same rate each year) for much of the last half of this century should have tipped scientists off that something was fundamentally wrong with our understanding of this greenhouse gas.

The methane increase began to slow back around 1982. Yet the IPCC, in its 1996 report, projected a constant rate of increase. By 1992, the growth of atmospheric methane dropped to *zero*. True to Kuhn's prediction that scientists are wedded to the prevailing "wisdom" (such as it is), scientists resorted to increasingly ornate and bizarre explanations. Among others, they included the effects of dust from the big explosion of the Mt. Pinatubo volcano in 1991, declines in fossil fuel emissions accompanying the economic slow-down of 1991–92, and drops in natural gas production in Russia.

Six years later, Pinatubo's ash was long gone and the global economy was booming. Russian pipes, however, were fixed up a bit.

We will bet that NCAR scientist David Schimel did not make a lot of friends when he admitted that assumptions about methane used to prepare the 1995 IPCC report were "based on an understanding of methane that was five to 15 years old." This fact is prima facie evidence of the perils of science by committee, which is how the IPCC operates.

E. J. Dlugokencky et al. published an assessment of methane's history in *Nature* in 1998. They suggest that atmospheric levels may be reaching a stable point because the amount of methane removed from the atmosphere by chemical reactions now equals the amount being added. While methane emissions have been relatively constant, their removal processes have been increasing.

The cause of methane's sudden stabilization is not completely known, but there is no doubt that airplanes have helped. Like automobiles, aircraft emit nitrogen oxides (NOx) as a result of high-temperature combustion. At the surface, NOx cooks in sunlight to form low-level ozone (O_3), a pulmonary irritant (which also absorbs cancer-causing ultraviolet radiation). At flight level, the sun is much more intense (as frequent fliers know, almost all clouds are below jet cruising altitudes), and the propensity to form ozone is even greater.

Ozone breaks down ultimately to OH⁻, known as the hydroxyl radical, which is highly reactive with methane. So airplanes ultimately accelerate the removal of methane from the atmosphere. The relative growth of air traffic far outpaces that of any other form of combustion-dependent transportation, including the automobile, so it is one likely cause of methane's stabilization.

About 15 percent of the current greenhouse enhancement results from chlorofluorocarbons, refrigerants whose manufacture is being phased out as a result of their role in depleting stratospheric ozone under the Montreal Protocol, another United Nations treaty. CFC concentrations are now largely stabilized and will decline slowly during the next 15 to 20 years. It is a pretty sure thing that CFCs are involved in removal of some stratospheric ozone, but the scientific community's fear-mongering about ozone depletion and increased cancer-causing radiation led to a major embarrassment, detailed in the accompanying sidebar.

Truth: Lost in the Ozone

Although stratospheric ozone depletion and chlorofluorocarbons are only a small part of the global warming issue, the same scientific and social dynamics operate.

November 3, 1993. *Science*, the prestigious journal of the American Association for the Advancement of Science, issues an "Embargoed Advance Information" bulletin: "Higher UV [ultraviolet] Radiation Strongly Linked to Thinning Ozone." Because no one has yet seen the paper, such prepublication notices ensure that skeptics and/or critics will be few and far between.

The article, by J. B. Kerr and C. T. McElroy of Environment Canada (the Canadian equivalent of the U.S. Environmental Protection Agency), contained the remarkable claim that "spectral measurements of ultraviolet-B radiation [the causative agent for basal cell skin cancer] made at Toronto since 1989 indicate the intensity of light . . . has increased by 35 percent

(continued next page)

(continued)

per year in winter." Almost every news article, and hundreds were generated, featured a dumbed-down version of this theme. Canadian Peter Jennings of ABC News reported the story with his patented look of concern and astonishment.

The heart of the paper was a graph of UV-B radiation for different seasons. "Winter" (December to March, in their definition) data were given as circles, and earlier and later data as crosses. The study was five years in length, quite a short period for any environmental trend. Further, the authors stated that they had no data for winter 1991–92 because the measuring equipment had been out of service. That leaves only four years.

A close look at their graph reveals that indeed they did have data for much of that winter, but that they were mislabeled as spring data! Not only that, but also all the ultraviolet readings were very low for the "missing" year. When 1992 was included, the trend in radiation dropped by nearly half. But more important, it was clear from looking at the data that the entire "trend" was induced by *four days* at the end of the record in 1993.

Four days out of the 419 days in the entire study. That is like getting normal blood pressure readings for 415 consecutive days, followed by four days of skyrocketing readings during a week in which you lost your job, your kid got arrested for drugs, and your weather-forecasting cat died—and then concluding that the high readings of those last four days constitute a significant trend.

Everyone knows that extremely high or low data at the beginning or end of a record induce a spurious trend. Those four days saw the jet stream disturbance associated with the so-called "storm of the century," the March 14–18 cyclone that caused blizzards, floods, or tidal inundation all the way from Florida (where two inches of snow fell) to southern Canada. Many places in eastern North America set their all-time record low barometric pressures during this storm.

(continued next page)

(continued)

Disturbances of this type are known to result in low ozone readings, so it follows that an extremely large disturbance would result in exceedingly low ozone readings and consequently high ultraviolet radiation.

Michaels pointed these things out in an article *Science* published in May 1994. This time, there was no "Embargoed Advance Information," nor did *Science* heed the call in Michaels' original manuscript that the original Kerr and McElroy paper be withdrawn. (In fact, the reviewers insisted that he remove the call for withdrawal to make the paper acceptable.)

Apparently the truth does not matter, even when published in *Science*. Even though Kerr and McElroy's analysis was shown to be wrong, in its 1999 *Special Report on Aviation and the Global Atmosphere*, the UN Intergovernmental Panel on Climate Change cited their paper as strong evidence for increased ultraviolet radiation caused by ozone depletion.

The facts were clear on this case. The original paper included a mistake, yet no one acknowledged it. Perhaps worse was that the glaring error made it through the initial peer review process in the first place. It is hard to believe that no one would have picked up it prior to publication, unless, of course, he or she was not looking very hard.

"Now we have good data to point to," Sherwood Rowland told the Associated Press. Clearly Rowland, the retiring president of the American Association for the Advancement of Science (the publisher of *Science*), had seen the paper before it was published. He subsequently received the Nobel Prize for his work on ozone depletion. Did that qualify him as one of the reviewers who missed this glaring error? If Nobel Prize–winners do not look very carefully when presented with results alleging environmental catastrophe, then who on earth does?

Ignoring the chinks in the paradigm—this time a deficit of increased ultraviolet radiation as a result of ozone depletion—is exactly what Kuhn predicts most scientists will do: "Normal science, for example, often suppresses fundamental novelties because they are necessarily subversive of its basic commitments."

There are a host of other greenhouse gases, but they are minor compared with the "big three" of CO_2, CH_4, and CFCs. For the purpose of assessing their ability to enhance the greenhouse effect, it is a simple exercise to treat the non-CO_2 greenhouse gases as if they *were* CO_2: Just consider their relative efficiency in absorbing the warming radiation from the earth's surface and their expected lifetime. One molecule of methane, for instance, acts like nearly 10 of carbon dioxide, and CFCs are exceedingly efficient warmers, given their tiny concentration in the atmosphere.

Hence we can calculate that the *effective* concentration of CO_2 is much greater than 365 ppm. Some of the equivalencies between effective CO_2 and other greenhouse gases are pretty naked estimates; for example, we can only infer the lifetime of methane in the atmosphere from what is an obviously poorly understood chain of reactions. But it appears the effective concentration is somewhere around 450 ppm, or more than 60 percent of the way toward a doubling of the post–ice age background average of around 280 ppm.

That represents a tremendous change in the capacity of the lower atmosphere to absorb warming radiation. Recall that the current paradigm holds that if computer simulations of the atmosphere are, in the IPCC's words, "generally realistic," then there should be an obvious and dramatic warming of the entire troposphere—the bottom 45,000 feet of the atmosphere. But as chapter 5 shows in detail, more than 80 percent of this layer has not changed its temperature a lick in more than two decades—a time when the greenhouse gases were growing most rapidly.

Greenhouse Changes vs. the Sun

A convenient method for tracking greenhouse changes is to express them in terms of energy within the atmosphere.

Consider a 1,370-watt light bulb shining over the earth. That is the radiance of the sun. Clouds and bright surface features such as snow and ice reflect away about 30 percent of that 1,370 watts. At any given time, half the earth is in darkness, and the bright half receives direct light only where the sun is vertical. As a result the sun effectively illuminates only one-

(continued next page)

(continued)

fourth of the earth. With all factors in place (some of which we do not know with much accuracy), about 342 watts remain to warm the surface.

The earth must radiate back out to space the same amount of energy it receives—otherwise the temperature rises. Changing the greenhouse effect reduces the effective outgoing radiation in the lower atmosphere, which causes the lower layers to warm up. Because of the combined effect of all the greenhouse enhancers, this change, called "radiative forcing," is thought to be around 2.5 watts per square meter by now.

That seems like a small amount compared with the 342 watts that bathe the earth's surface, but it is not. Consider the earth's temperature, not in degrees Celsius or Fahrenheit, but in Kelvin temperature, which is the number of Celsius degrees above absolute zero. The current value is about 289°Kelvin. The overall net incoming radiation of 342 watts works out to a little more than 1 degree (Celsius or Kelvin) per watt. While the matter is hardly that simple, changing the net radiation at the surface by 2.5 watts is not that small a matter.

How Much More Change?

Figure 3.3, taken from the 1995 report of the United Nations Intergovernmental Panel on Climate Change, shows five different scenarios for effective carbon dioxide concentration over the course of the next century. The reason the IPCC calls them scenarios rather than forecasts is that scientists do not want historians to tag them as having goofed. Anyone can make a scenario—all that are needed are a pen and a sheet of paper—but a forecast implies some specific knowledge about how a complicated system will work.

Does anyone really know what our primary energy source will be in 2050? How about in 2100? Does anyone know what our usage of energy will be?

The answers are obviously no, no, and no. Our fuel may change to nuclear fission, fusion, stay as fossil fuel, become fuel cells, or more likely, turn into something over the course of the next 100 years that no one has yet imagined. As a mental exercise, consider

Figure 3.3
IPCC SCENARIOS FOR FUTURE CARBON DIOXIDE CONCENTRATION

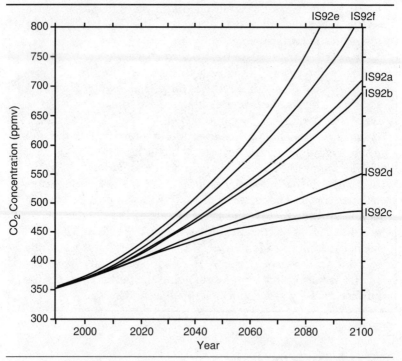

NOTE: The large scatter or discrepancy is indicative of the tremendous uncertainty that surrounds this problem.

100 years ago. Who would have had the following vision: Horses will be abandoned, waterpower left to the hauling of freight. Passenger ships and passenger trains will all but disappear; instead, people will fly around in aluminum tubes drinking wine bottled thousands of miles away while a box in their pocket plays Beethoven's 9th Symphony (something few 19th-century people had ever heard, owing to the lack of radio and rapid transportation). Those aluminum tubes will fly over cornfields yielding up to 200 bushels per acre (in 1900, 20 was pretty good) and over acres and acres of green fields planted with something called a soybean that can be woven into an ersatz meat or crushed to power a diesel engine (whatever that is) that runs giant motorized farm implements over a no-till field.

Anyone who made such a prediction would no doubt have been labeled one of a "small band of optimists" who fight against a mainstream of realists who anticipated nothing of the sort.

The point is that no one can correctly imagine the world of 100 years from now. It therefore becomes wise to abandon "scenarios" and stop at some point we are pretty certain will be reached, but beyond which we have little knowledge. Many scientists and policymakers who have looked at this issue think an effective CO_2 concentration of around 700 ppm (compared with the current 450) is about all we can be certain of, whatever "certain" means.

Remember that a mere 30 years ago many people were "certain" that massive population increases would overwhelm the earth. It was their calculations that drove energy use forecasts for today. The Club of Rome, in the 1971 computer simulation *The Limits to Growth*, employed those types of assumptions when they predicted we would be virtually out of petroleum and choking on pollution by now. Rome Clubbers Dennis and Donella Meadows produced a 1993 update, *Beyond the Limits*, that says the same thing, also predicting disaster 20 years hence. Paradigms die hard, but books threatening imminent destruction apparently sell well.

At any rate, an effective carbon dioxide concentration of 700 ppm seems an appropriate benchmark at which we can expect things to reach in the 21st century. How might this change affect the atmosphere? To answer this question, we must first examine what the atmosphere does.

The Circulation of the Atmosphere

Warm air and cold air behave differently. Warm air is more buoyant and holds more water than cold air. Warm air rises, cold air sinks. Moist air cools at a lower rate than dry air, so rising warm air is even more buoyant when it is wet. This creates the largest of the atmosphere's "primary" circulations, which gird the entire middle half of the planet.

The Intertropical Convergence Zone, Trade Winds, and the Subtropical Deserts

The location on the planet that best combines warm and wet is the tropics. There, large and persistent thunderheads transport heat away from the surface. A thunderstorm requires an inflow (convergence) of surface moisture and an outflow (divergence) aloft. Without these it literally runs out of air and collapses.

The hottest region of the planet, largely delineated by these thunderstorm clusters, is the intertropical convergence zone (ITCZ) and, where it is prevalent, it largely defines the tropical rainforest. The air converging toward the ITCZ forms two bands that are global in extent. These are the northeasterly and southeasterly trade winds in the Northern and Southern Hemispheres, respectively.

Air diverges aloft, away from these hot thunderstorm towers. Having lost a great deal of precipitation to the rain forest, the descending air actually warms at a greater rate than it had cooled on ascent (since moist air cools at a lower rate than dry air, dry air must warm at a greater rate than moist air). The descending air piles up about 30 or so degrees of latitude north and south of the ITCZ, giving rise to the world's desert climates. Technically, these air masses are known as subtropical high-pressure systems, or "anticyclones." In turn, the low-altitude flow back from the subtropical highs to the ITCZ creates the trade winds.

The Jet Stream, Fronts, and Cyclones

Far to the north, over the snow cover of northern Canada and Siberia, the sun's angle, especially in winter, is either very low or negative (i.e., polar night). The earth keeps on radiating at wavelengths characteristic of its temperature, losing more and more heat to space. The surface temperature drops further and further, resulting in shallow pools of cold air—generally around 4,000 feet in depth—that form the cold high-pressure systems of winter. These are called polar anticyclones, or polar high-pressure areas. They are some of the coldest air masses on the planet, and they are bone dry. As we explained earlier, that means they have very little natural water-vapor greenhouse effect.

Between the polar and subtropical high-pressure systems is a zone of low pressure that separates tropical warmth from polar cold. At the surface of the earth, these are the fronts that bounce up and down on the evening weather program. The net movement of air forward (i.e., cold air moving southward into the warm region) is a cold front, and backward (north) is a warm front.

These movements are largely controlled by the middle-atmospheric expression of the boundary between warm and cold known as the jet stream. Motion—in the form of the jet's rapid winds, which can exceed 200 miles per hour—is the atmosphere's way of dissipating the energy difference between warm moist air and polar cold.

The motion in the jet stream varies. Sometimes it is very fast and forms fan-shaped structures that drag up the air from underneath, a process that creates a surface low-pressure system, or cyclone. The cyclone itself moves the frontal boundary between warm and cold. In the Northern Hemisphere, the wind on the left side of a cyclone is blowing from west and north, resulting in a cold front—the cold boundary advancing southward. On the right side of the storm the opposite occurs and the warm air advances northward as a warm front.

In general, the more powerful the jet stream is, the more that air underneath it moves upward. Think of a strong wind blowing over a chimney. When the wind over the chimney howls, the fire in the furnace burns more strongly because air is being drawn into the column. When there is little wind, the fire dies down.

In the atmosphere, upward motion equals storminess, and downward movement means fair weather. So when the jet stream is strong, in general, things are stormier.

It follows that the strength, or speed and dimensions, of the jet stream is predicated upon the temperature contrast between dry, polar cold and warm, tropical moistness. The greater the contrast, the more powerful the jet stream, everything else being equal (dangerous words in this business).

Tropical temperatures do not change much through the year for two reasons. First, solar radiation changes very little compared with the polar latitudes. At the equator the furthest the sun gets from zenith (straight up) at noon is 23 degrees—barely a quarter of the way down to the horizon. At that solar height, there's very little change in the huge amount of radiation coming down from day to day. But in high latitude and polar regions, the sun's angle varies from mid-sky to, poleward of Arctic and Antarctic Circles, below the horizon, and the solar radiation change from season to season is spectacular.

Further, the world's tropics are largely oceanic, and the temperature of water, particularly over a deep ocean, changes at a much slower rate than the temperature of land. (This peculiarity largely explains why we go to the beach for summer vacation.)

On a seasonal basis, this makes the jet stream much stronger in winter than in summer, as the tropic-to-polar temperature change is much larger in winter, thanks to the relative constancy of the low latitudes. That is why, in general, cold fronts are more powerful (dropping the temperature more) in the winter than in the summer and why there are more and stronger low-pressure systems in the winter.

A propensity for strong cyclones also exists during the transition from winter to spring. As the jet stream lifts off to the north at this time of the year, it often leaves behind some transient whirls known as "cut-off lows." These are pockets of cold air aloft that, now detached from the normal flow, migrate slowly over the increasingly warm land for a week or so. A cut-off system in May 1992 dumped 60 inches of snow in the higher elevations of western North Carolina amidst the blooming dogwood. An analogous feature occurs during the transition from fall to winter, though it is not so pronounced as the jet stream because the jet stream is moving south rather than north.

Anticyclones

High-pressure systems are known as "anticyclones." They occur wherever atmospheric motions force air to accumulate. There are two main types.

Air descending from the ITCZ bottlenecks around latitudes 30°N and S and appears in the form of the dry high-pressure systems that determine the world's great deserts. These air masses lost the vast majority of their moisture when they precipitated out from the ITCZ thunderstorms. Their extreme aridity makes them candidates for enhanced greenhouse warming, meaning that the land underneath will change from a desert into—well, a desert.

The higher barometric pressure recorded in an anticyclone simply means that there is more air present. Anticyclones are divided into air mass types such as tropical and polar. This latter category—the cold anticyclone—deserves considerable attention because it kills more people, from cold and exposure, than any other anticyclone. In the United States, for example, the Bureau of Vital Statistics shows that deaths attributable to cold air annually exceed deaths from warm air by a factor of around four (see chapter 9).

One reason cold anticyclones form is that cold air is denser than warm air and therefore tends to "pool" or sink to the bottom of the atmosphere. Another cause for their formation is the slowing of the westerly jet stream, owing either to physical barriers (such as the mountains of western North America, which are virtually perpendicular to the westerly jet) or to the internal dynamics of the jet stream that induce transient accelerations and decelerations. When the jet slows down, air piles up underneath, and an anticyclone forms.

Siberia is surrounded on three sides by mountains and on the fourth by the frozen Arctic Ocean. In winter, cold air sinks into the Siberian Basin, creating the strongest anticyclones on earth. Most high barometric pressure records emanate from there. Temperatures in the cold Siberian anticyclones regularly drop below $-40°C$ ($-40°F$). The air masses are generally less than 4,000 feet in depth; for this reason, temperatures actually rise with height in much of the Siberian winter, unlike the "normal" situation on the planet.

Northwestern North America is also home to some very cold winter anticyclones, where speed changes are forced on the jet stream by mountains and land–ocean differences. These anticyclones usually are not so cold as their Siberian cousins, with an average temperature about $-30°C$ ($-22°F$), but their lowest temperatures can be below $-50°C$ ($-58°F$). Occasionally the jet stream kicks one of the major anticyclones southeastward toward the eastern United States. In the Christmas 1983 anticyclone, around 40 people (no one ever knows exact numbers for cold or heat deaths) perished in South Carolina alone. Those are deadly air masses.

El Niño

Most midlevel and senior climatologists were trained that the tropical and polar regimes interacted to produce the jet stream, causing the jet stream to produce storms and that is about that. We now know there is another large circulation that provides a connection between the tropics and the jet stream: El Niño.

As Figure 3.4 shows, most of the time the northeasterly and southeasterly trades blow across the Pacific into the ITCZ, moving the water at the top of the tropical Pacific westward. The net movement away from the South American coast toward the western Pacific

Figure 3.4
SCHEMATIC EL NIÑO

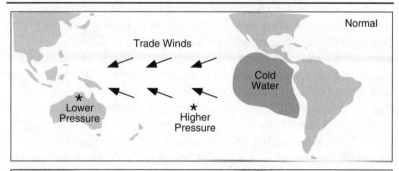

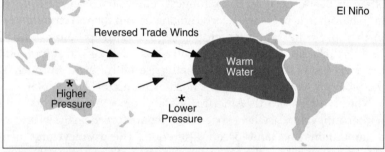

SOURCE: Henderson-Sellers and Robinson, 1990.

NOTE: Under El Niño conditions, the normal trade winds weaken or may even reverse, dramatically changing the temperature of the equatorial Pacific Ocean.

Ocean is a very real phenomenon, and as a result, the sea level in the Philippines is usually a foot or so higher than it is along the South American coast; during an El Niño, though, the flow across the Pacific is reduced, and the Philippine sea level drops. In El Niño's absence, the warm surface water that diverges from South America is replaced by upwelling colder water from the lower depths. That nutrient-rich water supports a tremendous growth of fish, mainly the anchovies that have been central to the Peruvian economy for millennia. For millennia, the culture of South America has adapted to El Niño.

The waters in the tropical eastern Pacific are generally a few degrees colder than those in the west. But the sun's migration through the sky imposes a seasonal cycle. In November or December, the trade winds often slack off for a few months, the cold upwelling is reduced, the anchovies go away, and Peruvian fishermen have a convenient vacation that just happens to fall around the Christmas holidays, hence the "El Niño" (the Child) moniker.

Sometimes the slackening of the trades lingers and becomes more pronounced. Or it may develop at a different time of the year. In that case the cold upwelling reduces dramatically, and warm water can actually spread across the ocean. This is the modern version of "El Niño," a *persistent* (as opposed to seasonal) slackening or even a reversal (from east to west) of the trade winds.

Under non–El Niño conditions, the trade winds move warm water westward across the Pacific and create a "warm pool" of water near the equator in the western part of the ocean. Given that warm air is more buoyant than cold air, the pool causes an enhanced area of thunderstorms in and near the ITCZ in the western Pacific. The storms represent a net movement of air—converged at the surface, sent aloft via buoyancy, and diverged aloft—that causes relatively low surface barometric pressures in the western Pacific. The opposite—cool conditions in the tropical east Pacific—results in sinking air, much of which originated over the warm pool, so that barometric pressures in the eastern Pacific tend to be higher than those in the west.

Discovered in the early 20th century, the phenomenon is called the Walker circulation. Its strength is given by the pressure difference between two points in the Pacific Ocean for which very long records have been maintained: Tahiti and Darwin, Australia, which is at the extreme north of that country. Known as the "Southern Oscillation Index" (SOI), the pressure difference is an excellent measure of whether or not an El Niño is occurring. When the SOI drops to or below the 17th percentile, climatologists like to officially pronounce the existence of El Niño. (Note that the statistical definition implies that there will be an El Niño, on average, 17 percent of the time. Hardly newsworthy, or so you would think.)

The weakening of the trade winds that accompanies El Niño reduces the upwelling of cold water off South America; in other words, El Niño prevents the cooling of the atmosphere that would occur in its absence. But the longer or stronger the El Niño, the more cold water is left to lurk just beneath the surface west of Peru and Chile.

When El Niño goes away and the normal trade winds return, an awful lot more cold water is available than there was, and temperatures plummet along the South American coast. The cold water spreads westward in a pattern that looks a lot like El Niño except that the warm anomalies are replaced by very cold conditions. This opposite effect has come to be known as La Niña, more accurately defined as the appearance of more cold water than average from South America westward.

There is a connection between El Niño and the westerly jet stream that was unknown before the era of weather satellites. We now know that under El Niño conditions, the volume of uplifted air in the eastern tropical Pacific is so massive that it is ultimately drawn into the midlatitude jet stream. The increase in mass movement causes the southern part of the jet stream to accelerate. Again invoking those dangerous words "everything else being equal," the result is an increase in upward motion or storminess.

A nice theory. In principle it must be right. But the atmosphere is a fickle thing. Scientists are quite good at theory and logic but oftentimes do not like to be disturbed by data and reality. In chapter 8, we take a closer look at the relationship between El Niño and weather in light of global warming forecasts. One thing that emerges is that El Niño correlates very little with overall changes in the United States—despite some "dynamic" arguments that attempt to relate La Niña to Midwestern drought. The arguments may have an internal logic, but the overall statistical association turns out to be profoundly weak.

El Niño/La Niña cycles are as natural as prunes. But their existence was known only to a very few climatologists and mariners until 30 years ago. It would seem from the media coverage of the 1997–98 El Niño event that some new and terrible thing had happened to the climate. Hardly. Just because scientists discover something or because we, as taxpayers, shell out millions of dollars to research something, does *not* mean that something new has happened. Chemicals existed before chemists; DNA existed before its discovery provoked a Nobel Prize; and El Niño ebbed and flowed long before the first climatologist was born.

It is reasonable, instead, to assume that Homo sapiens and just about every other species is adapted to El Niño, in one way or another. That is how evolution works—it takes advantage of environmental or biological changes and those that do not adapt get outcompeted by those that do. (It is no accident that many of the equations that drive population and community biology originated in academic economics.)

The point is worth a digression. Numerous studies, such as those of scientists in the U.S. National Oceanic and Atmospheric Administration (NOAA), demonstrate a very strong correlation between El Niño and drought in East Africa, which is where humans evolved. As the first of the sentient beings, our ancestors must have noticed years ago that the region they inhabited experienced prolonged droughts every 5 to 10 years or so. Which of our ancestors survived was determined in no small part by whether or not they had enough to eat. Perhaps the ideas of animal husbandry and purposeful planting of seeds to generate a surplus for bad times had something to do with the massive selective pressure that El Niño exerted on our ancestors.

It is hard not to poke fun at Al Gore here, wringing his hands at the El Niño Summit in October 1997 out in California. Everyone there who holds any position of environmental responsibility had surely noticed, without Gore's help, that years without El Niño tend to be pretty dry south of Santa Barbara, and years with El Niño tend to be wet. And further, Southern California has engineered itself around El Niño rains, as John McPhee noted so well in his brilliant essay in *The Control of Nature*.

Timothy Wirth, then under secretary of state for global affairs (read: global warming), also got in the act, telling Europeans that the El Niño was a symptom of global warming. Two months later, Wirth left State to head Ted Turner's newly minted billion-dollar foundation at the United Nations, devoted largely to spreading the word about the perils of global warming.

In California, Vice President Gore put forth the idea that the "terrible" El Niño might be caused by global warming; in doing so, he cited the highly speculative arguments of NCAR's Kevin Trenberth. (Trenberth had just testified before Congress on the matter.)

In response, Florida State University's Jim O'Brien, who has studied El Niño since before it was fashionable, said, "El Niños have been going on forever. We can trace them back in corals a thousand years, so they have nothing to do with global warming, or anything like that. I just wanted to get that straight because there was a

47

meeting in Congress today. Some . . . kept saying that because this
is the biggest one, that it's due to global warming. I hate this stuff. . . .
There have been bigger ones. . . . We certainly can find bigger ones
in the last century."

Others disagree with Trenberth on this matter, too (though we
should note that Trenberth has done some very creditable work on
droughts vs. El Niño, which we address in our section on observed
vs. predicted circulation changes). Consider the University of Wisconsin's Reid Bryson, arguably the intellectual godfather of the modern notion that humans can change the climate. When asked if he
thought El Niño becomes more frequent with global warming, he
told Michaels:

> This is absolutely, flat-out wrong. Sandweiss [D. H. Sand-
> weiss, who wrote a 1996 article in *Science*] saw it years ago
> when he showed that in the last 5,000 years there has been
> cold upwelling along the Peruvian coast, broken by intervals
> known as El Niño. Before 5,000 years ago there was always
> warm water, so there was continuous El Niño during ice
> ages, and cold water during interglacials. In other words, El
> Niño–type occurrences with warm water along the coast go
> with ice ages, not with warming.

Sandweiss' archaeological record indeed shows that El Niños are
less common in warm times. Coral records published by R. G. Fairbanks in 1997 in the journal *Coral Reefs* show stronger El Niños when
the planet was relatively cool, about 300 years ago. That they have
increased a bit in the 20th century as the planet warmed slightly
may simply mean that the strength of an El Niño does not relate to
the temperature, and that, as Bryson notes, El Niños may simply be
more common in climates like our ice-age prone era.

The 1997–98 El Niño deserves a bit more attention, however.
California Gov. Pete Wilson estimated that it caused about $500
million in damage in his state. (Incidentally, that amount is only a
quarter of the state's price tag for the big 1982–83 El Niño.) According
to the Federal Emergency Management Administration, when all
was said and done, weather-related damages in California in 1997–98
cost less than they had in the previous, non–El Niño winter.

Florida saw about $100 million in damage, mainly from a single
tornado outbreak that cannot directly be pinned on El Niño, anyway.
That tragedy killed 40 people because it occurred at night and in a

culture that eschews tornado warning sirens. Compare that with 1992's Hurricane Andrew, which cost $14 billion in Florida alone. The strengthened westerly jet that El Niño causes is known to weaken hurricanes in the Atlantic Basin.

According to the publication *Vital Statistics of the United States*, winter's weather-related death rates typically far exceed summer's. Most are a direct result of the cold—hypothermia, heart attacks from shoveling snow, pneumonia after slip-and-fall injuries. Federal computer models associate El Niño with warmer winters in the United States.

But the press takes little note of these things. Slowly freezing to death or expiring in a hospital just is not as newsworthy as getting spun to kingdom come by a tornado that someone says was related to El Niño. When the homeless die, as they often do in the cold, the news footage is somehow less heart-wrenching than the pictures of the bereaved outside a wrecked million-dollar home in Florida.

The truth is, El Niño suppresses hurricanes and induces a substantial financial benefit in the United States, where the average damage cost per year is now $5 billion. (See chapter 8 to learn whether or not hurricanes are becoming worse as the planet warms. Hint: No.) An intense or extreme hurricane (category 4 or 5 on the Saffir-Simpson scale, a bit of arcana that everyone who watches the Weather Channel knows) will produce about $25 billion in damage, depending where it hits. According to Colorado State University's Bill Gray—the man who can predict next year's hurricane season better than anyone in history—El Niño reduces the likelihood of this type of hurricane by at least one-third and sometimes more. The average frequency of these monsters' striking the United States (about one in seven years) works out to an El Niño savings of about $1.25 billion per year.

According to Energy Department statistics, the very warm winter of 1997–98 saw the demand for heating energy drop about 15 percent. People liked to finger the huge El Niño for this because the strengthened subtropical westerly winds associated with it cut off the normal Arctic air of winter and bathed the country in California-like bliss. The average heating cost per winter in America is around $50 billion, so El Niño directly saved about $7.5 billion there.

The reduced demand for heating oil resulted in a glut of the alternative refinery product: gasoline, which ran as low as 89 cents per gallon for much of the winter for regular grade. That resulted in the average car using about a dollar less in gas per week. Conservatively estimating that there are about 200 million cars in regular use, and that El Niño was responsible for half the price decline, gives us another $2 billion.

Of course a lower gas price significantly reduces the cost of transportation. The boom in transportation stocks—which outpaced the Dow Jones industrials by a factor of two that winter—increased personal wealth by several billion.

Then there is the vegetation, agricultural and otherwise. California became greener than a blackjack table (a nod here to fellow Cato fellow P. J. O'Rourke, who once referred to Uruguay as "greener than a crap table"). The abundance of rain in the Great Plains brought the price of wheat and other small grains to record lows, after adjusting for inflation. Count a few more billion for the consumer.

It is easy to demonstrate at least $15 billion in El Niño benefits in the United States, balanced by a tautly stretched $2 billion in costs. That was the estimate Michaels made in April 1998 in the *Los Angeles Times*. By September 1999, climatologist Stanley Changnon calculated a similar figure in the *Bulletin of the American Meteorological Society*. Al Gore conflates El Niño, global warming, and terrible things in the United States at his peril.

Primary and Secondary Circulations

So far we have been discussing mainly the primary circulations—the chief engines of the atmosphere. The late Speaker of the House of Representatives, Tip O'Neill, used to say, "All politics is local." So is all weather. Whether or not an El Niño is cooking means nothing unless it produces something tangible and destructive. In and of itself, a jet stream–driven cyclone is merely an area of low surface barometric pressure. But if that cyclone spins up a tornado-producing thunderstorm, people will surely notice. The secondary circulations that result from the prime motions of the atmosphere, then, are the currency of weather and climate.

Hurricanes. The hurricane, the most severe member of the family of objects known as tropical cyclones, is the single most destructive storm on earth. Even though they are called cyclones, hurricanes' lateral dimensions can be as small as one-tenth the size of a jet stream–induced cyclonic low-pressure system. And the core of destruction, usually in the strongest wind band surrounding the calm eye, is often no more than 30 miles wide.

These secondary circulations are quite powerful indeed. They can form as disturbances within the trade winds, or they can be initiated by a southward excursion of the westerly jet stream over very warm water. Hurricanes of the first type—trade wind disturbances—are, in general, the real death angels, whereas their jet stream–driven counterparts often produce more good than harm in the form of beneficial rains with only modest property loss.

Hurricanes converge the moisture of tropical oceans into huge pinwheels of thunderheads. As moisture condenses, heat is released to the environment; the core of a hurricane can be more than 11°C (20°F) warmer than the surrounding region. The increase of motion near the warm core draws ever-increasing amounts of tropical moisture inward, resulting in a classic positive feedback loop. Once a hurricane begins to power up, it can progress from a modest 75-mph storm (category 1 on the 1-to-5 Saffir-Simpson scale of hurricane intensity) to a 156-mph (category 5) killer in less than two days.

In 1999, Hurricane Bret grew from category 1 to category 4 in a remarkable 18 hours. When it made landfall north of Brownsville, Texas, it had the distinction of being the only category 4 or 5 storm to ever strike the United States (there have been 15 this century) and directly kill not one person. This is a tribute to the ability of affluent societies to protect themselves infrastructurally and warn the citizenry. (It also did not hurt that Bret hit on one of the few stretches of the U.S. East Coast that is not teeming with vacation homes.)

Only two category 5 storms have ever struck the U.S. coast. "The Labor Day Hurricane" of 1935 was a weak system when it exited the Bahamas but blew up to what was probably a 160-mph monster (no one knows the true highest winds in that storm) as it crossed the Florida Keys only 36 hours later. Hurricane Camille in 1969 took about two days to go from a modest storm to a category 5 before splashing into the Mississippi Gulf Coast with winds that were estimated at 193 mph, judging from onshore damage patterns.

Hurricanes feed on warm water. They are not likely to form if the sea surface temperature is less than 27°C (80°F), and they are not likely to maintain themselves, once formed, with temperatures below 25°C (77°F). The notion that human beings could warm the

51

globe's mean temperature would seem to naturally lead to the conclusion that this would produce more or more severe hurricanes. But plenty of warm ocean is available today to spin up hurricanes—plenty more, in fact, than hurricanes take advantage of. So it is unclear that adding to this area makes any difference whatsoever.

Further, hurricanes require more than just a warm ocean. As the air ascends skyward in the thunderstorm columns, it must be diverged from the top—otherwise the entire process will stagnate. (Imagine a fireplace, which must have a chimney with an open flue to draw properly.) This "outflow" mechanism is critical to hurricanes. Favorable outflow conditions are less common than sufficiently warm water is; some theoretical considerations indicate that favorable outflow conditions might become even less frequent with greenhouse warming.

El Niño suppresses hurricane activity in the crowded Atlantic Basin even as it enhances activity in the remote central Pacific. Those who claim global warming will increase El Niño frequency or intensity, which, in turn, will cause a net change toward more severe or frequent hurricane destruction, are on very thin ice indeed. Atlantic storms are costly, owing to the huge valuations for vacation property along the U.S. East Coast and in the Caribbean; but central Pacific storms find little land and few people to run into. Those hurricanes that make it to crowded, affluent places such as Hong Kong or Japan kill very few people, many fewer than a storm of the same strength in, say, Vietnam, proving once again that wealth and infrastructure trump climate almost every time.

Tornadoes. Tornadic winds are the most destructive movements of air on earth. Tornadoes occur in severe thunderstorms, which themselves consist of massive updrafts of warm air and downdrafts of cool air. The updraft can be seen on hot days in the form of rapidly rising cumulus clouds that ultimately become thunderheads; the downdraft is the waft of cool air that usually precedes the rain by 15 minutes or so.

When the updraft is spun by a disturbance in the westerly jet stream and stretched by the rising motion, the result is a concentrated whirlwind of accelerated winds that may be a half-mile across. At the center of this vortex the winds rise rapidly as the barometric pressure drops. If the drop is large enough, moisture condenses and a typical tornado funnel forms. The corollary—that weaker tornadoes do not show a classic funnel—is also true, which means there are a lot more tornadoes out there than people once thought.

That is about to become common knowledge. The new National Weather Service Doppler radars, designated WSD-88, measure both precipitation and the velocity of precipitation. Older radars simply painted a still picture. And what these WSD-88s are finding are more and more tornado vortices in thunderstorms that show rotary motion but are not strong enough to condense a solid funnel. As a result, more and more specific tornado warnings are being generated on stormy afternoons, but, in reality, there is no real change. There is not more weather—just better weather sensors.

With regard to the greenhouse effect, two competing factors should affect tornadic thunderstorms. The first concerns vertical motion, and the second relates to the required spin.

The temperature contrast between the surface and the upper atmosphere determines the amount of upward motion, of buoyancy, required to produce a thunderstorm. Everything else being equal, warming the surface without warming the rest of the atmosphere should increase buoyancy. Greenhouse models indicate a general warming including the surface and the layers of the atmosphere dominated by thunderstorms. But this is not at all what has happened in the last quarter of this century. Instead, the layer near surface has warmed a bit while the rest of the troposphere has not. The net result of the surface warming accompanied by little or no change in the rest of the weather zone is a slight increase in buoyancy.

With regard to spin, remember that the temperature contrast between the pole and the tropics determines the jet stream's strength. That contrast is greatest in winter and spring and least in the summer. Because changing the greenhouse effect reduces the temperature contrast, a greenhouse change may reduce the amount of energy required to spin up tornadoes.

Overall, it is probably pretty close to a draw. A slight increase in buoyancy (which was *not* forecast by the greenhouse models) argues for slightly more powerful storms, but a decrease in the spin caused by a slightly weakened jet stream argues for the opposite.

Floods and Droughts. The 1995 IPCC report contains several remarkable passages, but perhaps none is as strange as this one:

> Warmer temperatures will lead to . . . prospects for more severe droughts and/or floods in some places and less severe droughts and/or floods in others.

The industry press often singles out this statement because it blames virtually every type of weather anomaly on global warming—even the prospect of weather that is more normal than average (i.e., "less severe droughts and/or floods").

According to Michael MacCracken, Director of the Interagency Office of the U.S. Global Change Research Program—a congressionally mandated office whose purpose, among others, is to assist in the cutting of large portions of the $2 billion annual pie on global environmental change—the IPCC was attempting to communicate here that the cycling of water through the atmosphere becomes faster on a warmer planet and that this increases rainfall intensity. At the same time, increased evaporation caused by warmed temperatures allows the surface of the planet to dry more rapidly, increasing the likelihood of drought. We think those conditions seem a bit exclusive—most places where it is raining heavily are not experiencing a drought! Thus the "and/or."

At the same time, some areas of the planet experience their rainiest conditions as a result of annual warming—monsoonal India, for example. May temperatures in New Delhi regularly reach 43°C (110°F). The warm land of central India, along with the warming Himalayan mountain surface to the north, draws the tropical moisture of the Indian Ocean inward, which creates a monsoon. So maybe warm conditions will increase rainfall in this "other" place.

The fact of the matter is, both the IPCC's statement and MacCracken's defense of it are based upon the paradigm that the computer models of the atmosphere are, in the IPCC's words, "generally realistic." Those models predict all the elements of climate described in this chapter will change in potentially important ways. Whether or not the forecasts of gloom and doom are true depends on whether the computer models are reliable. The nature of the models requires exploration.

4. Modeling the Earth's Climate

When confronted with the idea of a changed greenhouse effect, just what is the right tool for assessing future climate? Two schools of thought have emerged. One clings to computer models as the only way to predict the future; the other holds that, since we have been enhancing the greenhouse effect for more than 100 years, nature has already given us the answer.

This bifurcation is common in all science—climatology, psychology, economics, and so on: Empiricists and dynamicists have always quarreled, as have liberals and conservatives, young and old, and men and women. Both, however, do have their points.

In their university courses, the authors have always taught that this dichotomy is unfortunate and serves to the detriment of our ultimate understanding. The two factions should snipe less and talk more. Nature has something to say about models and models have something to say about nature.

The Empirical School

The empirical school is generally identified with three institutions: the University of Delaware, the University of Virginia, and Arizona State University. Some of the academic papers coming from this group examine the long history of certain climate variables (temperature, rainfall, jet stream positions, and so forth) and compare them with what we would be expect in a world with an enhanced greenhouse effect; others may also compare these histories to the GCMs' projections.

The empiricists believe that there has been plenty of time—100 years or so—for nature to declare the amount and type of response that climate displays to the changed greenhouse effect. In the last chapter we demonstrated that, because of the panoply of greenhouse emissions, we have already gone more than 60 percent of the way toward doubling the natural carbon dioxide greenhouse effect in the atmosphere.

That is a lot of change. It means that the atmosphere effectively has a carbon dioxide greenhouse effect that has not been seen' for at least 10 million years, and probably longer, although there is plenty of evidence that during the Cretaceous era (when dinosaurs ran the place), it was several times higher than it was right before humans took over. Writing in *Science* in 1996, C. I. Mora et al. found a concentration of around 4,670 ppm, or more than 10 times today's value, about 300 million years ago—during a period when the earth was probably the greenest it has ever been.

At any rate, the empirical school holds that the changes in the greenhouse effect have been sufficient to have established at least a substantial hint of nature's course. By simply matching the observed global temperature history of the last 100 years with changes in the greenhouse effect (and making no other major assumptions), it is simple to calculate that the planet should warm about 1.4°C (2.5°F) with an effective doubling of atmospheric carbon dioxide concentrations.

The argument for the empirical school stems from the continual increase in greenhouse changes over the past century. It would be shocking if there were no warming. But the amount certainly seems to be tolerable, and, given society's propensity to change its energy and way of life on the century time scale, most people in the empirical school do not see global warming as a major concern.

In a way, the empiricists, too, are "modeling" the climate. Their model merely states that we have seen an exponential increase in the greenhouse effect concurrent with modest change in temperature.

General Circulation Models

There are hosts of pressing climatological problems that simply cannot be solved by examining historical records. For example, we have known for more than a century that, for much of the last two million years, massive glaciers have covered the earth, extending as far south as Chicago, and that these glaciers have waxed and waned several times during that period. Even when ice ages were at their height, the glacial front made substantial north and south movements.

Why do ice ages occur? The question is certainly pertinent from a practical point of view. We are throwing greenhouse gases and dusty particulate emissions in the atmosphere, which could serve to warm and cool things, respectively. It would be nice to have an accurate computer model of this process. We also know that we have radically altered much of the preexisting surface color of the earth, also with potential climatic consequences. Again, it would be nice to have a model.

The ice age (and other climate change) problems attract additional attention because the earth seems perilously close to another freeze-up. Given humans' ability to alter the basic building blocks of climate—the amount of solar radiation reaching the surface, the rate at which that radiation escapes out to space, and the energy absorption of the surface—it seems a good idea to explore how these changes might affect the climate.

Relatively simple equations of climate exist that can say, on the whole, how the temperature of the whole earth–atmosphere continuum could change given the range of possible human impacts on those building blocks. But such a result is meaningless because it has no geographic resolution. The earth is not a point, and we do not all live everywhere at once. Determining how climate changes both horizontally and vertically as humans go about their merry economic activity is a more difficult enterprise.

Moreover, the climatic regime in which we and the plants we depend upon thrive consists of a number of variables distributed throughout the day and the year. Altering the composition of the atmosphere can change the distribution of warming energy within the day. In a simple example, consider that much of the dust produced by civilization is hygroscopic, meaning that water molecules tend to adhere to it. The wetter it is, the more they adhere. As they grow in size, the particles increasingly reflect and scatter the sun's radiation, while also serving as the basis for increased cloudiness. All of this means that the particulates human activity produces can alter both the mean temperature (by reflecting away radiation that would normally have served to warm the surface) and the daily energy balance. Clouds cool the day, but at night their substantial water-vapor greenhouse effect produces warming.

That is just a simple example. If the clouds are deep enough, they will produce precipitation. That precipitation falls on the surface as either snow, which is white and will reflect away the sun's warming radiation, or rain, which may serve to darken a soil surface, causing the earth to absorb more heat. The moisture in the air is then recycled in the form of evaporation, but evaporation itself cools the environment.

This myriad of interrelated problems, describable at least in part by simultaneous differential equations, is meant for computers. Indeed, some of the original computers were invented specifically to solve the same equations that determine the processes that give us our day-to-day weather. Computers, using equations whose details are not completely known, move heat, mass, and moisture around both horizontally and vertically as the greenhouse effect and other climate factors change. The resulting general circulation model represents the integration of many of the important processes.

Horizontal and vertical resolutions are crucial. The number of computations required to run a GCM varies geometrically with the number of points at which all of the equations are solved. So a model with twice as high a resolution requires four times as much computation. This is a daunting problem since the world's largest supercomputers are already employed running GCMs.

Resolution varies and has improved with time. The earliest GCMs, 20 or more years ago, simulated the climate using the concentric latitude bands around the planet and only a few vertical layers. By the time of the Framework Convention on Climate Change, horizontal resolutions of 3° x 5° latitude/longitude (36,000 square miles on this part of the globe) became common. Which means that when the computer created rain or snow, it was always over an area, in the middle latitudes, of a region the size of Iowa. On the Weather Channel, the dimensions of summer thunderstorms, which produce the vast majority of midlatitude rain at that time of the year, Iowa-size storms are uncommon compared with, say, Chicago-size ones.

The National Aeronautic and Space Administration climate model, which gained more early press than any other because of the commitment of its operator, James Hansen, to publicizing it and gaining political attention, actually resolved at a massive 8° x 10° degrees latitude/longitude during that era. That is about *five* Iowas, or 192,000 square miles.

In global terms, newer models really have not changed their horizontal resolution very much (though many are down to 2° x 4° degrees, a slight improvement), but they have increased the number of vertical layers. So we are still stuck with nearly Iowan thunderstorms even though one might think we are better at estimating how climate changes vertically. This latter, alas, turns out not to be the case.

The "Parameterization" Problem

"Climate" is the result of the interaction between incoming solar radiation and the amount that the earth absorbs or retains. This simple definition belies some complexity.

Knowing the albedo, or reflectivity, of the surface is vitally important to determining how much radiation the earth absorbs. It is possible personally to appreciate the power of albedo by wearing black on a sunny, warm day. Switching to white changes the amount of solar energy absorbed from around 90 percent (in black clothes) to 10 percent (in very white)—and can mean the difference between heat exhaustion and relative comfort in a desert environment.

The earth's albedo is around 28 percent, meaning that it reflects a little more than one quarter of the radiation it receives. Changing the albedo only a few percentage points changes the surface temperature by several degrees.

That hypothetical summer thunderstorm is about as white as a cloud gets, reflecting up to 90 percent of the sun's radiation in its cumulus core. A feathery cirrus cloud "anvil" can reflect about 50 percent; the average for all clouds is around 60 percent. In a model that resolves at 3° x 5° degrees latitude, each and every summer rainfall event—even a passing shower that might affect one square mile—covers the whole horizontal grid spacing of 36,000 square miles! In Iowa, the reflectivity of a cloud-free surface during summer is around 20 percent, so the computer may be reflecting away three times as much of the sun's radiation (60 percent vs. 20 percent) by making every square-mile shower into a grid-size rainstorm.

These errors add up. In fact, unless they are artificially stabilized with "flux adjustments" (that is scientific jargon for "fudge factors"), and "parameterizations" (scaling the results or the input with other fudge factors), most climate models *must* run away into an unrealistic climate. To assume that all of these errors would exactly cancel each other out—creating a stable computer-generated climate—is simply fantasy. Consequently, the NCAR model was known to "drift" into a successively colder earth, and the model from the United Kingdom Meteorological Office piled up too much heat on the wrong side of the Gulf Stream. Hansen's NASA model was rumored to melt the polar ice cap without increasing carbon dioxide.

Albedo is not constant. People—and nature—change it dramatically. Reflectivity changes naturally every time there is a forest fire. It even depends on time of day. Consider how much more of the sun's rays the ocean reflects around sunrise and sundown (which makes sunset at the beach so attractive) than at midday. And a windswept ocean, whipped up with whitecaps, reflects more than a calm sea. If the greenhouse effect changes, does the ocean's reflectivity also change, owing to changes in wind patterns? By how much?

This last question is the most important one in climate science. Without knowing "how much"—without knowing the correct value of a climate input parameter—no one can ever estimate accurately how the climate output can or will change. Without knowing *how much* humans have altered the surface of the earth, for example, we can only guess. Those guesses are "parameterizations."

Consider the parameterization of clouds. Clouds are concentrated water vapor—so much so that many are "supersaturated" and actually have more moisture than they can hold over time. That moisture turns into rain. Rain develops when tiny cloud droplets run into each other and coalesce: When their size is large enough, they fall to earth.

The point is that clouds exert a substantial regional greenhouse effect because of all their water vapor, an effect we can observe almost every winter. On a clear, calm night, temperatures drop precipitously after sunset, as cold air is usually very dry and has little natural greenhouse effect. But when a cloudbank rolls in under such conditions, the temperature stops falling. If it is early enough in the evening and the surface has lost only a small portion of the day's heating, a nighttime cloudbank can completely arrest the usual fall in temperature.

Clearly, any systematic error a GCM makes in the formation of clouds or rain (and the grid size *ensures* such errors will occur) will induce substantial miscalculations of regional (and therefore global) temperature.

Even more difficult is cloud type. Low, puffy clouds, such as stratus and the familiar fair-weather cumulus, are so bright that their ability to reflect the sun's radiation from above exceeds their ability to trap heat (via the greenhouse mechanism) from below.

Consequently, low-level clouds are net coolers of the surface temperature, and increasing their coverage results in a drop in global mean surface temperature.

Above roughly 20,000 feet in the midlatitudes, most clouds are ice crystals. The prime example is the feathery cirrus cloud. The optical thinness of these clouds (you can usually see the sun through them) means that they are reflecting not nearly so much radiation as the lower cumulus clouds; and, in the case of thin cirrus, the effect is a slight net warming as their greenhouse effect overtakes their albedo effect.

What happens if we increase the overall cloudiness of the atmosphere? The answer was unknown until 1988, when *Science* published the satellite studies of Ramanathan, who found that the net effect of clouds is a slight planetary cooling.

That GCMs produce clouds that are thousands of square miles in extent guarantees unrealism. And unrealism at any vertical level means that all the other levels are similarly suspect, since the atmosphere behaves like a stirred fluid. If the temperature in the bottom of a stirred cup is misestimated, then so is the temperature of the whole cup.

It is therefore wholly unlikely that GCMs could produce a correct or even realistic mix of cirrus and cumulus clouds, one of which warms and the other of which cools the surface.

What, we might ask, are GCMs good for? In their original conception, they were designed to explore the complicated atmospheric dynamics that determine our ever-changing climate. They were not intended, in their initial incarnations, to serve as vehicles to chart global environmental and economic policy. Instead, they are "what if?" research and teaching tools, and as such they have been informative.

For example, only a GCM can provide quantitative guidance on how the atmospheric circulation—particularly the jet stream—would change as the earth's mountain ranges rise and fall or as continents drift. Thanks to GCMs, part of the ice age mystery has apparently been solved: According to research published by Ruddiman in *Nature* in 1998, the rapid rise of the Himalayas, as the Indian subcontinent continues to barge into Asia, induced an increasing tendency in the jet stream over North America that allows ice ages to develop.

This kind of gross calculation—relating geological-scale events such as mountain-building to the ice ages—is really about as accurate as a GCM can be. As for specific predictions about the climate of a given point (see Figure 4.1 in insert), GCMs were and are largely inadequate. Even the IPCC rates its confidence in such predictions as "low." Asking GCMs how the local climate changes for relatively small changes in the atmospheric greenhouse effect (which is all that human beings could ever induce on the atmosphere compared with the 33°C [59°F] natural greenhouse effect resulting largely from planetary water vapor) is asking too much, too soon.

Unfortunately, as far as plants and animals are concerned, much climate is local and divorced from the primary circulations. Redwood trees do not care whether there is a jet stream or an El Niño, which a GCM may be able to simulate. They just care that it is cool and that there is a lot of fog in July on the hillsides where they grow, things a GCM cannot project.

GCMs have constantly evolved since the landmark early work of Manabe of the Department of Commerce's Geophysical Fluid Dynamics Laboratory in Princeton, New Jersey, in 1975. The first models did not have continents, which were added around 1982. The first models assumed that the ocean was a shallow swamp; by 1990, atmospheric models were "coupled" to an ocean that had both shallow and deep layers. By 1990, the IPCC had written that their results were "generally realistic" when fed the current mix of greenhouse gases.

Although there are literally dozens of GCMs now in existence, only five were heavily used at that time; they predicted, on the average, that the surface would warm 4.2°C (7.6°F) if atmospheric carbon dioxide doubled. The models were variously referred to as NCAR (for U.S. National Center for Atmospheric Research), GFDL (for the U.S. Department of Commerce's Geophysical Fluid Dynamics Laboratory), NASA-GISS (for NASA's Goddard Institute for Space Studies), UKMO (for United Kingdom Meteorological Office; now the "Hadley Centre" model), and OSU (for Oregon State University; now University of Illinois). Besides a net global warming, they predicted that high latitudes warm more than the tropics and that winter warms more than summer.

Ubiquitous Linearity

Figure 4.2 (insert) shows that GCMs characteristically predict a *linear* greenhouse warming. At first this might seem odd, given that the increase in greenhouse gases in the last chapter is slightly exponential (meaning that it forms an upward-pointing curve). In large part, the linear warming forecast results from an exponential change in the greenhouse effect superimposed upon an earth whose surface temperature will "lag" the greenhouse "forcing" by several decades.

This linearity occurs because the earth does not respond in lock step to changes in the greenhouse effect. The ocean, for example, has a considerable lag; if there were no lag, sea and air temperatures would be the same. Put another way, water has a greater "heat capacity" than land—referring to the ratio of the amount of heat absorbed by a substance to its observed temperature rise.

The massive depth of water on this planet has something to do with that. Because it is deep, the ocean is slow to warm and then "catches up." Together with the exponential increase in greenhouse gases, these two concepts can sum to what is nearly a straight line.

Figure 4.2 shows clearly the ubiquitous nature of linear forecasts of temperature rise. It seems quite remarkable that there can be such disparity in slopes (linear change) but such agreement as to the functional form (straight line) in so many GCMs.

This constancy of trend yields one important "law" of global warming: *If the climate models are right, then once a linear warming trend is established, that trend will continue at the same rate.* That the models give different slopes demonstrates that they really do not know how much it will warm. Once that trend is established, however, there should be little need to further question the amount of expected warming.

There is another reason that temperature warms at a constant rate while carbon dioxide increases exponentially. The response of temperature to carbon dioxide eventually "damps out." Assume there is no CO_2 in the atmosphere and therefore no CO_2 greenhouse warming. Adding the first increment—say, a million tons—will (after allowing for thermal lag) create the first increment of warming. Adding a second increment of exactly the same size will not produce quite as much warming as the first one. By the time we get to the umpteenth increment (and we are hardly there yet—there is evidence that atmospheric CO_2 has been as much as *17 times higher* during periods when the planet teemed with life) the result is little additional warming: In short, it levels out.

Mathematically speaking, the response of the atmosphere to a given greenhouse gas is logarithmic. At higher values, this function shows lower increases for a given perturbation than at low values.

Coupling this response with the earth's thermal lag gives another linearity: The warming for, say, quadrupling carbon dioxide is very close to a doubling of the temperature change produced by doubling carbon dioxide.

Our next chapter compares model predictions with reality; but the continued stream of alterations and "improvements" through the 1990s indicates that there must have been some substantial systemic errors that were not generally being trumpeted to the public. Put simply, here is the core error: *No GCM has ever succeeded in creating a troposphere (the bottom 40,000 feet of the atmosphere) that behaves at all like the observed data of the last quarter of the 20th century.*

Michaels noted a similar problem in a manuscript he sent to *Science* in 1990, which the editors returned with a terse note saying that the modeling community was well aware of this problem so there was therefore no need to publish his paper (an expanded version appeared in the *Bulletin of the American Meteorological Society* in 1992).

In other words, while the United Nations was promoting the paradigm that the models were "generally realistic" and using them as the basis for sweeping policy recommendations that could gravely harm United States prosperity, the models were in fact making massive errors that the IPCC was loath to note.

After all, if the GCMs were calculating unrealistic average temperatures for the troposphere, how could they make predictions about the future? In other words, if a GCM calculates that the earth is currently several degrees warmer than it actually *is*, what logical device allows it to make a forecast of future warming?

GCM forecasts of future climate change were never pure calculations of surface temperature. Instead, the *model-calculated* temperature without the changed greenhouse effect is subtracted from the *model-calculated* temperature with the changed greenhouse effect. The purpose of this exercise is to subtract out all of the systemic errors!

The Sulfate Fix

Dividing much of his time between NCAR and the University of East Anglia in Great Britain, where his small army constructs global temperature histories, Thomas M. L. Wigley is one of the most enigmatic figures in this entire issue. Wigley was the doctoral student of Sir Hubert Lamb, whose magnificent 1972 treatise, *Climate: Past, Present and Future,* remains the benchmark history of the world's weather and its interaction with civilization.

Lamb was an early champion of the notions that climate could change in ways that are significant to humanity and that people might have something to do with causing it to change. Recognizing the enormous political and economic implications of his views (and knowing full well their limitations), Lamb never translated his findings into social activism.

For his part, Lamb was more concerned about global cooling than global warming. He paid special study to volcanoes, which cool surface temperature, developing a history of volcanic eruptions and their influence on climate. Meanwhile his colleague and friend Reid Bryson, at the University of Wisconsin, pointed out that the activities of human beings tended to raise a lot of dust and to create finely divided atmospheric aerosols that he argued would also cool the climate. Much of this dust, Bryson said, was agricultural, and he once calculated that if every human on the planet merely kicked one tablespoon of dust skyward on a windy day, it would be enough to initiate a cooling of ice-age magnitude.

It was Wigley who took the climate issue into the realm of social activism where his mentor Hubert Lamb would not tread. Wigley, a federal employee, is also under contract with the Pew Foundation, a philanthropic organization that is very vocal in its support of the Framework Convention on Climate Change. He is possessed of an adept mind and keen political instincts. For example, he was one of the first of the global warming activist-scientists to recognize that if the models were fatally impugned, there would be no real reason for the world to enact the sweeping programs he believes are necessary. In a recent issue of *Earth* magazine, Wigley wrote that much more than the current climate treaty will be required to sufficiently address global warming.

Wigley realized early on that the climate models were predicting too much warming, and wrote about it in a 1987 issue of *Climate Monitor.* By 1990, Wigley had adapted the Lamb/Bryson cooling mechanism to the global warming models, offering little credit to its originators. He reasoned that if some type of aerosol cooling were compromising greenhouse warming, then the GCM output could be more easily reconciled with reality (Figure 4.3 in insert). (Again, as Kuhn predicted, a scientist strives to bring observed data in line with the prevailing paradigm.)

Of course, no "real" carbon dioxide greenhouse effect in climate models causes warming and no "real" particulate aerosol cools them. After all, GCMs are only series of equations that attempt to mimic the theoretical implications of the inputs. In the case of aerosol, because that dust tends to reflect sunlight, the amount of solar radiation in the model that directly warms the surface is reduced. There is no doubt that performing the operation will reduce the warming in a computer model, whether or not it is happening in nature.

Wigley's analyses rely heavily on the Lawrence Livermore National Laboratory (LLNL) Model, which received increasing attention after the publication of the 1990 IPCC report that relied instead more heavily on the five models noted earlier. Its first major publication was Karl Taylor and Joyce Penner's 1994 article in *Nature*. The paper was a tribute to Kuhn's notion that scientists behave increasingly strangely as paradigms become challenged.

Everyone from the IPCC to the most fervent Doubting Thomas on the subject of climate disaster agrees that the greenhouse effect has been changed by the addition of industrial gases to the tune of around 2.5 watts per square meter, as we explained in the last chapter.

But that is not what is in the LLNL model as published by Taylor and Penner. Though it uses the wrong change in the greenhouse effect, the LLNL model appears to produce an interesting result. When fed only the (wrong) greenhouse increases, it produces temperature changes for today that were clearly wrong and largely too high, but when the temperature figures were modified with sulfate aerosol, a much more realistic picture emerged, with some evidence even indicating that eastern North America should show little if any warming by now (a fact that has always bedeviled global warming proponents).

Reproduced here (Table 4.1) is an accompanying data table from the Taylor–Penner publication. As is quite plain, the model uses only half of the known changes in the greenhouse effect (1.26 vs. 2.5). Further, the sulfate "cooling" is given as 0.9 watts per square meter, globally averaged, which is at the very high end of estimates

Table 4.1
RADIATIVE FORCING FROM GREENHOUSE GASES

Global Average	Change in Forcing (watts per square meter)
Present-day total positive charge	1.26
Present-day sulfate	−0.95
Combined CO_2 and sulfate	0.31

SOURCE: Taylor and Penner, 1994.

NOTE: The known total positive charge is 2.5 watts, or twice as much.

given in the scientific literature. The net difference of 0.36 watts (1.26-0.90) produces considerable warming when there should be very little.

When the "right" number for greenhouse forcing, 2.5 watts, is used, even with their probably unrealistic sulfate cooling, the model produces a tremendous amount of warming, as shown in our accompanying figure. When the correct greenhouse effect is put in, the projected warming for today's greenhouse effect over the Southern Ocean is an outrageous 9°C (16.2°F). Observed changes have been about 0.3°C (0.5°F) in this region. The predicted warming is 3,300 percent of the observed value.

One might argue that this is the "equilibrium" warming that occurs after all the ocean's thermal lag is taken up. Being 3,300 percent below what is currently observed suggests it would take roughly 1,000 years to show up in full.

In April 1996, Joyce Penner presented the LLNL model at a seminar held in one of the House of Representatives office buildings. The seminar was sponsored by what was then the U.S. Global Change Research Program headed by Michael MacCracken, who himself was "on loan" to the USGCRP Washington office from LLNL. USG-CRP holds frequent seminars in the House, almost all of which bear the sheen of the doomsaying cant so as to provide cover for the administration's (mainly Gore's) views and to embarrass the Republican Congress. Only one "mainstream skeptic," John Christy of the University of Alabama, has ever been allowed to present a seminar in this venue, and only if his prominent adversary (and doctoral

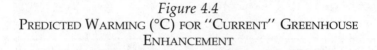

Figure 4.4
PREDICTED WARMING (°C) FOR "CURRENT" GREENHOUSE
ENHANCEMENT

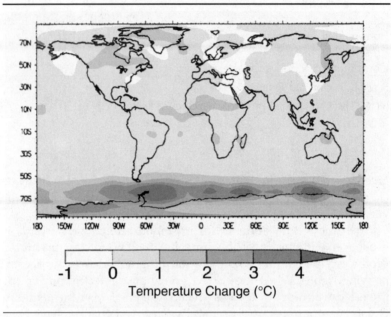

NOTE: This model uses one-half of the known change in the greenhouse effect and a very large sulfate cooling effect.

thesis adviser) Kevin Trenberth would speak immediately afterward. (The USGCRP is so scared of an unchallenged view against its interest, that it will simply not permit one. Your tax dollars at work!)

Penner was asked at the seminar why the LLNL model used only one-half of the known greenhouse changes. She replied that they were only looking at changes caused by carbon dioxide alone.

That reply was absurd. The title of the 1994 paper was "Climate System Response to Aerosols and Greenhouse Gases . . ." (emphasis added). Carbon dioxide is only one, singular gas that humans contribute to the greenhouse effect (Figure 4.4).

Why would LLNL be interested in simulating only one half of the greenhouse changes (roughly half is from carbon dioxide, the rest being from other anthropogenerated greenhouse gases)? Is this

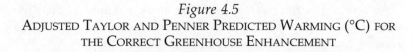

Figure 4.5
ADJUSTED TAYLOR AND PENNER PREDICTED WARMING (°C) FOR
THE CORRECT GREENHOUSE ENHANCEMENT

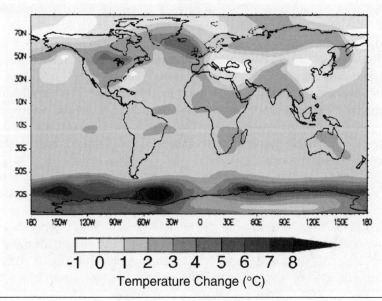

NOTE: This figure is the same as Figure 4.4, only adjusted upward with a realistic greenhouse effect change. This is clearly an unrealistic result, and strongly suggests only half the known change was used to begin with.

what the taxpayers get for their billions? Clearly, the problem was that, when fed the proper greenhouse effect, the LLNL model was exceedingly warm.

The speaker before Penner was the University of Washington's Robert Charlson, who showed how important the sulfate "rescue" to the overheated climate models was. "If we do not narrow our understanding of the uncertainty on the aerosols," Charlson said, "our understanding of the entire greenhouse effect is at stake."

By then, the IPCC had recognized that something major was amiss with the GCMs, models that a mere six years earlier they had called "generally realistic." One month after Penner's Washington seminar, the IPCC released its 1996 report, which included a statement admitting that the climate models used in its first report tended to predict a "greater mean warming than has been observed."

Soon most laboratories had a new generation of models, this time employing some type of "countering" of global warming by sulfate cooling. Not surprisingly, every model became cooler.

The time history of the Hadley Centre model, published in the *Journal of Climate* in 1995, produces a warming of approximately 1.8°C (2.9°F) to date *without* sulfates. What sulfates largely do is delay warming, which delays our crossing the 1.8°C threshold until somewhere around 2060. Note that, even with sulfates, the warming trend remains linear rather than exponential.

Unfortunately, major problems remain. For example, the UKMO model requires large "flux adjustments" for oceanic heat because it has the unfortunate tendency to improperly transfer warmth across the Gulf Stream, the most important point of heat transfer in the entire Atlantic Ocean.

NCAR's new "Community Climate Model 3" (CCM3) does not contain oceanic flux adjustments. When run with sulfate aerosols and a 1 percent per year increase in effective CO_2, it produces only 2.0°C (3.6°F) of warming in the next 100 years. Before this model, wrote Richard Kerr, *Science* magazine's global warming maven, flux adjustments were widespread. "Climate modelers have been 'cheating' for so long it's almost become respectable," he observed.

Even the IPCC does not assume 1 percent per year; instead it uses a mean figure of 0.63 percent. Starting GCMs off at a 1 percent per year effective increase in CO_2 brings them from late 19th-century concentrations to today's value somewhere back around 1934! So it might not be a bad idea to adjust the NCAR CCM3 output and place in a more reasonable term for the CO_2 increase. This is simple to do by taking advantage of the nearly linear relationship between CO_2 and time-dependent warming that exists in GCMs, and we have performed this exercise in Figure 4.6. Warming over the next 100 years drops to around 1.4°C (2.5°F). As previously noted, this is the same value obtained by merely extrapolating the warming of the last 50 years forward. Is nature trying to tell the modelers something?

But 1.4°C (2.5°F) of warming in a hundred years is just not enough to scare the citizenry. And there is no way that the American people will agree to dramatic rises in energy prices if that is all the warming we will see globally in the next century.

During the hot summer of 1998, public opinion polls began to pick up that Americans, though they might say global warming is

Figure 4.6
WARMING TREND OVER 120 YEARS PREDICTED BY NCAR MODEL.

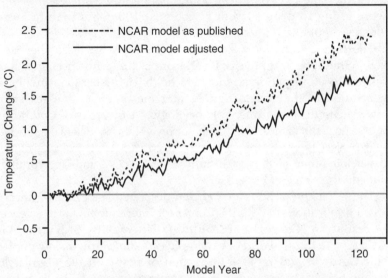

NOTE: As originally published (Kerr, 1997a), this non flux-adjusted GCM had an effective carbon dioxide increase of 1 percent per year—an amount the observations do not support. We have adjusted this output to reflect a more realistic, if still conservative, increase of 0.7 percent per year (in fact, the observations for the past two decades show even less of an increase— 0.4 percent per year).

"important," indeed were not sufficiently exercised about the issue to do very much. Environmentalists trumpeted victory (when they should have been terrorized) by one poll that found a majority of Americans would be willing to pay 25 cents more per gallon of gasoline if that would stop global warming. But the pollsters did not stop at a quarter increase! The rest of the story is that few were willing to pay even 50 cents per gallon more.

There just was not enough fear in America. The lowered estimates of warming were beginning to leak out. Standard rhetoric was that "as models improved, they tended to predict less warming."

A new organization, A Consortium for the Analysis of Climate Impact Assessments (ACACIA), was accordingly set up, headed by Wigley. ACACIA funds were used to generate "new" sulfate scenarios for the next 100 years. Suddenly a nonrefereed report, with Wigley's authorship, appeared, saying that it was going to warm faster than the CCM3 indicated after all, because sulfate emissions were going to drop further in the next 100 years than he previously thought.

The sulfate scenarios were prepared by ACACIA researcher Steven J. Smith. He is a physics Ph.D. from UCLA with specialties not related to atmospheric science. He holds two jobs: One involves generation of the sulfate figures, and the other is as a research associate in the Department of Sociology at University of Colorado, where he is the principal investigator on a National Science Foundation project to study the careers of scientists. That project's home page notes that a very large number of Ph.D.s wind up working outside their area of expertise.

The sulfate projections are four self-described "storylines," unrefereed publications of CIESIN (Center for International Earth Science Information Network) at Columbia University. Storyline A1 "describes a future world of very rapid economic growth, low population growth and rapid introduction of new and more efficient technology.... In this world people pursue personal wealth rather than environmental quality." And so on. Storyline B2 "includes concerted efforts for rapid technology development, dematerialization of the economy, and improving equity."

Obviously these are nothing but politicized fairy tales now driving a climate model. They only have one thing in common. In every scenario, the cooling effect from sulfates is dramatically reduced.

This fairy tale–based science was then published by the Pew Foundation, in the nonrefereed booklet authored by Wigley. Pew, via its Pew Center on Global Climate Change, is nakedly advocating for passage of stringent limitations on greenhouse gas emissions (i.e., energy use). In its press release concerning Wigley's booklet, Pew stated that those scenarios would be incorporated into the IPCC's next full assessment of climate change.

This is a remarkable story. ACACIA, headed by a federal employee, uses four "storyline" scenarios for the future that run the political gamut from "pursuing wealth rather than environmental quality" to "dematerialization" and "improving equity." This is never-never world. In the real one, high per capita income is largely correlated with environmental quality and longevity, while low income is associated with degradation and death.

These "storylines" are then to be used by the IPCC, which will hold them up as the "consensus of scientists." No doubt someone will then testify to a Senate committee that anyone who disagrees is "not mainstream," and that these scenarios are reason to pass onerous legislation.

The absurdity of this approach is patent. Wealth cleans the environment and poverty degrades it. The lousiest air on the planet is in Africa, a pretty poor place, followed by China, not exactly wealthy. China is the largest single national source of sulfate aerosols, because of its heavy reliance on coal. The storylines are predicting the energy behavior of China over the next 100 years when we could not even figure that they were stealing our nuclear weapon designs for 10 years.

By July 8, 1999, when his Pew study was the lead item on the MSNBC nightly news, Wigley had input the "storyline" scenarios— not into the new CCM3 ("Community Climate Model") at NCAR— but into a small model easily run on a large calculator because it contains few differential equations, treats the earth very simply, and has no thunderstorms. He chose this path, he said, because his model could "run quickly," allowing these scenarios to be inspected with dispatch. By pursuing "personal wealth rather than environmental quality," Wigley had pushed up next century's warming to 2.9°C (5.2°F), or 33 percent beyond the CCM3.

Ironically, two months earlier, on April 14, Wigley had issued a press release based upon CCM3, which featured only 2.0°(3.6°F) of warming. As noted, that number should be further reduced by 30 percent because of a likely overestimation of the rate of greenhouse enhancement. Who is right here, the guy with the hand calculator, under contract to the Pew Foundation, or the guy with NCAR's Cray X-MP supercomputer, funded by the federal government?

Such is the power of sulfate aerosols. Take them out and the world warms faster. But if, in fact, they are not very important anyway, the world simply will not warm that much. Indeed, the sulfate card is what holds up the artifice of disastrous global warming.

5. Has the Earth Warmed?

With a reluctant nod to the Clinton administration, this seemingly simple question depends upon what the definitions of "has" and "earth" are. Depending upon that definition, we can answer yes, no, or do not know. Why? Because the earth's temperature is hardly constant. The earth has warmed and cooled for billions of years, and the current ice-age regime is one of the most variable periods in that inconstant history. It is an astounding fact to behold that this planet is around five billion years old but has likely seen large areas of glacial ice for only around one half of 1 percent (a crude estimate at best) of its total history.

This Ice Age is hardly over. We are merely between glacial phases; indeed, we are due for a reglaciation, if the history of the last million years or so is any guide.

The Ice Age Earth is a planet whose temperature fluctuates wildly, and we just happen to be here during that era, cheerily emitting compounds into the atmosphere that are themselves known to change the surface temperature. Finding the human fingerprint on an atmosphere at the height of chaos is a daunting task indeed.

So what do "has" and "earth" mean? If they mean the surface of the planet in the last 100 years, there is doubtless a warming. If we mean the "free troposphere" (the atmosphere largely removed from surface disturbance, all the way up to the stratosphere) in the last two decades, the answer is that there is no net change. Further, what we see depends upon what we use for measurement.

At first glance, it would seem easy to determine if the planetary temperature has been increasing over the past century. Many weather records throughout the world extend for more than a century. In theory, we should be able to assemble those records, check for trends, and easily determine whether or not the world is warming. This exercise has been carried out and, based on the results, the world is indeed warming.

Many global temperature records are available from the thermometer network of the world, including the NASA temperature history.

But the most popular and widely used record has been developed and maintained by Phil Jones of the Climate Research Unit at the University of East Anglia. That data set is based on the records of several thousand land-based stations and millions of weather observations taken at sea. He converts the monthly station observations into 5° latitude by 5° longitude grid-box data, and all values are expressed as deviations (anomalies) from a reference period defined as 1961 to 1990. The grid box anomalies may then be areally averaged for each hemisphere, and the two hemispheric values are averaged to determine the estimate of global temperature.

The most popular version of Jones's history is the one employed by the United Nations Intergovernmental Panel on Climate Change (IPCC). That record blends land and sea temperatures to create a global average. The blending technique is hardly straightforward, as land-based and sea-based temperatures are not exactly commensurate. Consider that most land temperature records originate from standard "shelters" a few feet above the ground, while most ocean temperatures originate from some type of ship-based platform. The differences between a thermometer shelter and a ship are rather obvious. Further, the method of taking oceanic temperature has changed over the century, from canvas buckets thrown overboard and retrieved to engine intake tube measurements. Reconciling the changes has required a lot of assumptions and guesswork.

These considerations aside, another more fearsome problem emerges. A long-term climatic history is, by definition, made up of long-term stations. Why and where were most weather stations established? The long-term records almost all originate at some type of commercial center. In other words, cities have a way of growing up around their weather stations.

This situation induces a slight but real warming trend that has nothing to do with the "true" temperature. Sometimes this "urban effect" is recognized, and the "official" station is moved to a more rural location. In Chicago, for example, the "official" station was first moved from the central city out to Midway Airport. As one of the nation's busiest airports during the piston-powered era, Midway attracted a lot of commerce, and the city eventually grew out to and around it. In the 1960s, the "urbanization" of the Midway record became obvious, so the "official" station became O'Hare field, then a largely underused concrete elephant amidst fertile cornfields. Anyone who travels knows it is now a very different environment whose

Figure 5.1
Schematic Effect of Urbanization on Temperature Trends

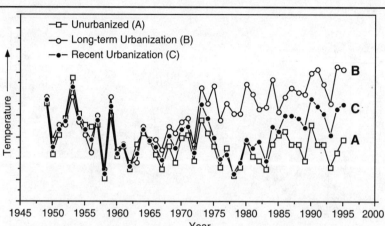

NOTE: Records A, B, and C all show the same variation from year to year, but B has an upward trend, indicating urbanization, and it is removed from global temperature histories. Record C begins to warm only in recent years, so it is erroneously retained in the global history.

urban characteristics do not differ much from Midway or the rest of the city.

Jones and others have attempted to deal with the problem as best they can. Two nearby temperature records are compared year-to-year. If they go up and down together, but one of them has a warming trend that does not appear in the other, the former record is assumed to be suffering from urban warming and is removed from the history.

So far, so good. In Figure 5.1, these appear as "A," an unurbanized station, and "B," an urbanized one that clearly shows a trend. Year to year, though, the records bounce up and down together.

Despite the wishes of the many, cities have a way of sprawling into the surrounding countryside, and airports have a way of adding concrete. So a relatively pristine station such as the 1960s-era O'Hare has changed from a patch of stone embedded in cornfields, to one surrounded completely by masonry, buildings, and automobiles.

The problem is that a station like O'Hare does not begin to show the signs of urbanization until fairly late. "C" in our example represents such a station. The ups-and-downs match very well through

Figure 5.2
NORTHERN HEMISPHERE ANNUAL TEMPERATURE HISTORY, 1900–1999

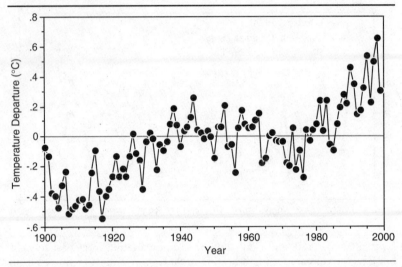

SOURCE: IPCC, 1995 and updates.

NOTE: This is the 20th-century Northern Hemisphere surface temperature history the IPCC uses.

the entire record, but the warming only shows in the last decade or so. Any statistical tests to isolate this are likely to fail because the year-to-year variation swamps the very real but short-term urban warming.

Population continues to grow, though not as fast as some had warned. The largest increments are in the most recent years, so the probability that a weather station "goes urban" increases significantly near the end of its history—precisely at the point for which we have no objective mechanism for isolating the effect.

The bottom line is that Jones and the IPCC have largely removed the urban effect when it dominates a temperature record for many decades. But for the last 10 or 15 years, no known method exists to get rid of it. The urban effect is here, and it will grow exponentially.

Plotting the IPCC temperatures leaves little doubt that the earth's mean surface temperature has warmed during this century (Figures 5.2 and 5.3). The degree to which this is consistent with forecasts of *human* induced change is highly debatable, as shown in the next

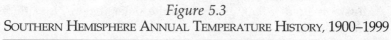

Figure 5.3
SOUTHERN HEMISPHERE ANNUAL TEMPERATURE HISTORY, 1900–1999

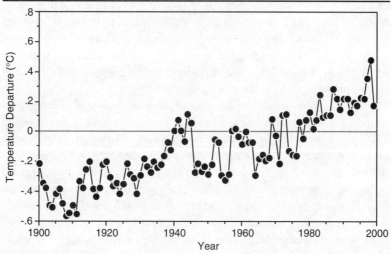

SOURCE: IPCC, 1995 and updates.

NOTE: This is the 20th-century Southern Hemisphere temperature history the IPCC users.

chapter. The linear trend in the entire 1890 to 1995 period is 0.6°C (1.1°F); In the Northern Hemisphere, where there is much more data than in the sparsely settled (and mainly water) Southern Hemisphere, temperatures rose about 0.4°C (0.7°F) from 1900 to the mid-1930s. They then fell about 0.3°C (0.5°F) through 1975. Since 1975, surface readings have warmed and now stand a mere 0.2°C (0.4°F) above values typical of the 1930s, or six decades ago. The 13 warmest years on record all occurred from 1980 onward, and the 15 coldest years all occurred before 1920.

But the urbanization effect is difficult to remove from the end of the record. What is more, these thermometric surface air temperature estimates are fraught with other problems as well, including the lack of data in remote and oceanic areas, changes in the network over the past century, changes in instruments and observation practices, and microclimatic changes near the weather equipment, such as a growing tree near a weather station.

All GCMs predict that, away from the polar regions, the atmosphere above the surface warms more than the surface, especially in the tropics. Often known as the "free troposphere," this zone, from 5,000 to 30,000 feet, is largely independent of the earth's varied surface and should behave in a much smoother fashion, responding nicely to the increase in greenhouse gases. Above the troposphere, as noted in the last chapter, a general cooling is predicted for the stratosphere.

We are fortunate to have three records of free tropospheric temperature. One is a satellite-based time series that extends from January 1979 to the present, originally published by Roy Spencer of NASA and John Christy of the University of Alabama in *Science* magazine in 1990 and updated monthly. The satellite senses the temperature by measuring microwave emissions from molecular oxygen in the lower atmosphere. Microwaves are able to penetrate the atmosphere with little attenuation, and the amount of energy the satellites receive is directly proportional to the temperature there. On a global scale, the accuracy of the satellite temperatures is thought to be ±0.01°C (0.02°F). The instrument package, called the Microwave Sounding Unit (MSU), rides an orbit inclined to the pole and therefore covers virtually the entire planet twice a day.

A plot of the resultant global temperature anomalies (Figure 5.4) certainly looks different from the surface record. The satellite-based global temperatures reveal a statistically significant warming of 0.05°C (0.09°F) per decade in the Northern Hemisphere; over the same time period the near-surface air temperatures warmed by 0.15°C (0.27°F) per decade—three times the satellite-observed amount. In the Southern Hemisphere the satellite finds no significant change, although surface records show a warming.

The slight warming trend in Northern Hemisphere and global MSU temperature is purely driven by the heat of the 1998 El Niño working its way out to space. Take that year away and there is no trend. Figure 5.5 shows another representation of the MSU data, in *daily* readings expressed as the departure from the long-term average since January 1, 1997. It is obvious that temperatures peaked in March 1998 and have been on a rather steady decline since then as El Niño waned.

The reason we know that 1998's heat has to do much more with El Niño than with the greenhouse effect has to do with stratospheric

Figure 5.4
MONTHLY TEMPERATURE HISTORIES FROM THE MSU SATELLITE SENSORS

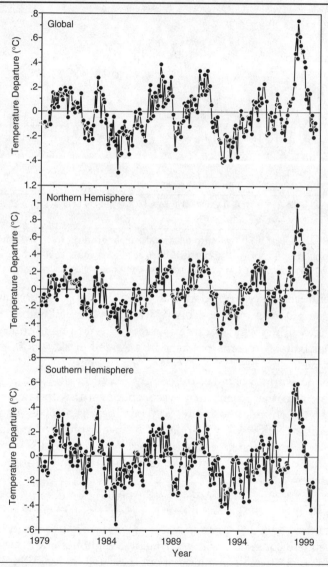

NOTE: These figures show global (top), Northern Hemisphere (middle), and Southern Hemisphere (bottom) temperature histories from the MSU satellite sensors.

Figure 5.5
GLOBAL DAILY SATELLITE TEMPERATURES

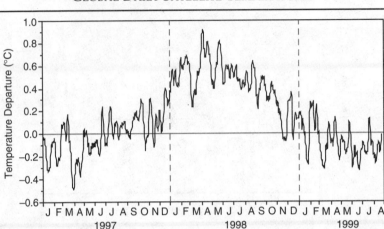

NOTE: This plot of daily satellite temperatures beginning in 1997 shows the pronounced spike from the big El Niño. It also reveals that temperatures returned to below the long-term average by early 1999.

temperatures. As we noted in chapter 3, changing the greenhouse effect should induce a tropospheric warming coupled with a stratospheric cooling. That cooling is everywhere to be found in the weather balloon records published by Angell et al., whose global record begins in 1958 and extends to the present.

El Niño warmth behaves differently, however. It is a true pulse of heat from the oceans that wafts spaceward, working its way through the entire atmosphere. So when there is a spike in surface temperature followed by a spike in the stratosphere, that must be El Niño and not the greenhouse effect. With greenhouse changes, warming of the troposphere is accompanied by cooling of the stratosphere.

Our plot (Figure 5.6) of Angell's stratospheric temperatures clearly indicates that 1998 showed the warmest stratospheric temperatures in a decade and that there is a clear overall negative trend in the stratosphere consistent with a changed greenhouse effect.

"We've got to do something about the satellite," said the Union of Concerned Scientists' Harold Ris at a White House global warming pep rally held in October 1997, prior to the UN meeting in Kyoto designed to make the climate treaty "legally binding."

Figure 5.6
LOWER STRATOSPHERIC TEMPERATURES

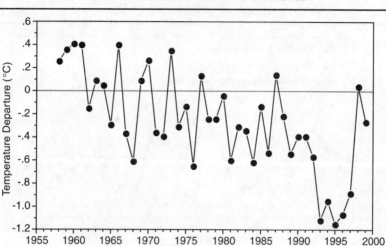

SOURCE: Angell, 1994, and updates.

NOTE: The lower stratospheric temperatures archived by the Department of Commerce's James Angell show a steady decline since they began in 1958 that is highly suggestive of an expected greenhouse signal. The spike in 1998 is the warmth of El Niño superimposed on the expected cooling.

In turn, California rocket scientist Frank Wentz calculated that the slight drag that the very thin atmosphere exerts upon the MSU satellites should induce a tiny decay in the orbit that would result in their drifting ever closer to the earth. The resultant smaller "footprint" for the MSU sensor, left uncorrected, would induce a small but spurious negative (cooling) trend in the data. Rather than wait for the customary peer review, Wentz made sure his finding went straight to the top—to none other than Vice President Al Gore, whose staff let every environmental journalist in the nation know that the satellite data were, in their words, about to be "discredited."

Wentz calculated that the orbital decay of the satellite would induce an artificial cooling trend of 0.12°C (0.22°F) per decade. When coupled with the satellite-observed (1979–97) cooling of 0.04°C (0.07°F) per decade, this left a "true" warming of 0.08°C (0.14°F) per decade. This is still more than twice as low as the mean value

Figure 5.7
MSU SATELLITE, WEATHER BALLOON TEMPERATURE, AND WEATHER BALLOON BAROMETER READINGS

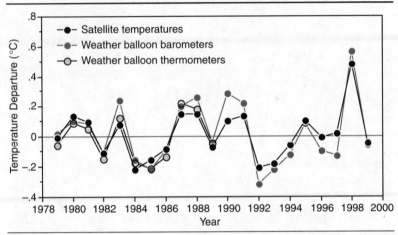

NOTE: These three independent measures all show the same year-to-year variation and have no warming trend (except the now-departed El Niño spike of 1998).

predicted by the sulfate-greenhouse models described in the last chapter, but it represented a small victory—at least it was the right sign!

Note that our satellite temperatures do not display much warming at all. That is because Wentz neglected two other drifts in the satellite, known as east–west and time-of-day. This corrected version of the satellite data has been accepted for publication in the refereed journal *Atmospheric and Oceanic Technology* and will appear in early 2000; this book uses the corrected data.

Wentz completely ignored that there were two other measures of free tropospheric temperature that meant the satellite was still right. Twice every day, weather balloons are launched to provide a vertical profile of the atmosphere in order to initialize (start) the computer-generated daily weather forecast models. The instruments are all known and standardized. The balloons carry electronic temperature and pressure sensors (barometers), and their altitudes are carefully checked.

As Figure 5.7 shows, temperatures measured by weather balloons between 5,000 and 30,000 feet line up perfectly with the temperatures sensed by the satellites. So Wentz, in essence, is arguing that somehow the weather balloons and satellites are making the very same errors in temperature measurement—day after day, for more than 7,000 days. The temperature record we use is the same one Oort published in the journal *Climate Dynamics* in 1989 and the same one used by federal climatologists in their comparisons of modeled and observed temperatures.

Ironically, the satellites are so good that they were ultimately used to correct an error in some Australasian weather balloon readings that developed when the supplier for the temperature sensor was changed. This has led some people to argue that the two sets are now so confused that they have lost their independence. But that is not the case for another record we use—this one from Angell of the U.S. Department of Commerce—which consists of temperatures calculated barometrically.

Thermometers measure temperature directly. But another way to measure the temperature takes advantage of this equation, which a few readers may remember from college chemistry: $PV = nRT$.

This is the "ideal gas law," which states that if you know the pressure and volume (P and V) of a gas, and you know how many molecules there are (n), then using the constant, R, you can calculate the temperature, T. The weather balloon measures P (pressure) and its height is known. The ascent path can be considered constant, which means that height times the path gives volume. Standard atmospheric tables give n, and R is the same everywhere, known as the "universal gas constant."

So the balloons' barometers provide an independent check of the temperature. In addition to the satellite and the temperature data between 5,000 and 30,000 feet, our graph shows the temperature in that layer calculated from Angell's barometric pressure readings. All three readings go up and down in unison for each of the 20 years that make up the entire record, and there is no warming trend (except the big El Niño spike in 1998).

The satellite begins in 1979, but the global weather balloon record extends back to 1958. In the 5,000-to-30,000-foot slice, the balloon record shows a linear warming trend of 0.09°C (0.16°F) per decade. The surface temperature trend is virtually the same at 0.10°C (0.18°F) per decade. This is about four times less than the greenhouse-only models of the last chapter predicted and is a bit more than two times what the sulfate-greenhouse models forecast.

How can we reconcile the obvious disparity between the last two decades (where there is no trend, after allowing for 1998's El Niño) and the entire (1958–present) weather balloon record, which does show a trend? How can we reconcile the fact that surface thermometers since 1979 show a warming trend of 0.15°C (0.27°F) per decade while the satellites and weather balloons show nothing?

The National Research Council (NRC) attempted to do this in a report, *Reconciling Observations of Global Temperature Change*, released in January, 2000. Indeed, the NRC concluded that the difference between the satellite, balloon and surface temperatures was real, *and that it revealed a serious flaw in the climate models*. In their words:

> "It is clear from the foregoing that reconciling the discrepancy between the global-mean trends in temperature is not simply a matter of deciding which of them is correct or determining the ideal "compromise" between them. In the long term, it will require *major advances* [emphasis added] in the ability to interpret and model the subtle variations in the vertical temperature profile of the lower atmosphere."

NRC panel chairman John Wallace, of University of Washington, told the *Washington Post* on January 13, "There really is a difference between temperatures at the two levels that we don't understand." In fact, most climate models predict that the temperatures measured by the satellites and the balloons should be rising faster than those at the surface.

The NRC report is a watershed and underscores the arguments made throughout this book. But what it doesn't say nonetheless reveals the some remarkable behavior by the IPCC. In their 1995 report, the "Policymakers Summary" contains not one mention of the word "satellite". Many of the authors of that report were also on the NRC panel. Where is the explanation of this very blatant attempt to mislead those who shape global warming policy?

In fact, the combined behavior of the surface and satellite records underscores two great mysteries of the atmosphere.

Mystery No. 1: What if there were a sudden and dramatic warming, and no one noticed?

Given that the weather balloon record from 5,000 to 30,000 feet shows no warming in the last 21 years, but does have an overall trend in it when extended back to its 1958 beginning, you might

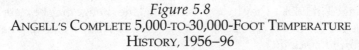

Figure 5.8
ANGELL'S COMPLETE 5,000-TO-30,000-FOOT TEMPERATURE HISTORY, 1956–96

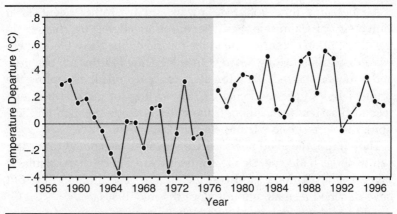

NOTE: An examination of the entire history reveals a distinct jump in 1976–77 that explains all the warming trend in the entire record.

SOURCE: Angell et al., 1994.

conclude that there had to be a pretty healthy warming trend in the first two decades, from 1958–78. Not true. For reasons that are largely unknown, all of the warming in the balloon record is compressed into one year—roughly the 12 months surrounding January 1, 1977. Figure 5.8 shows the annual average temperature in the entire balloon record in two segments, 1958–76 and 1977–96. In both sets of data there is no warming trend, but the offset between the two averages is about 0.35°C (0.63°F).

Amazingly, this sudden climate change was not even *noticed* by the scientific community until 1990! And the science editors of the nation's daily newspapers did not catch on until *1998*, 22 years after it occurred, when Thomas Guilderson and Daniel Schrag, writing in *Science*, described what is now known as "The 1976 Pacific Climate Shift."

By measuring the nutrient uptake history from Pacific corals (which reflect annual temperature in a fashion somewhat analogous to the more familiar tree-ring histories), Guilderson and Schrag found that "the vertical structure of the eastern Tropical Pacific changed in 1976." They hypothesized that the change may be responsible for the relatively strong and frequent El Niño events that have occurred since then.

At the same time, John McGowan and two other scientists at University of California, San Diego, found interesting changes in the distribution of marine life in the Pacific. In the early 1980s, there were disastrous fishing seasons in the Pacific Northwest. Seabirds and sea lions starved. Plankton abundance dropped, disrupting the primary link in the food chain. Cheerleading for global warming–related disasters (even though the authors made no such implication), the wire services trumpeted the bad news.

They forgot the good news. McGowan et al. also noted that Alaskan fisheries had "spectacular shifts upward" in catch, particularly salmon. Pollack, hake, and cod yields also rose dramatically, but have declined a small amount since the mid-1980s.

And it did not look much like global warming, either. McGowan et al. wrote that all of this had happened before. "Many fish and invertebrates were found well north of their usual range in the summer of 1926," they wrote. California fishery production dropped dramatically in 1960, following the 1958–60 El Niño, which they described as "one of the largest in the past 80 years."

What McGowan did not mention was Steinbeck's 1945 novel *Cannery Row* about the destitution of Monterey, California, resulting from a dramatic decline in sardine fishery after a similar shift in ocean temperature.

A final note: There has never been an adequate explanation for the sudden warming of the midatmosphere in 1976–77.

Mystery #2: Why does not warming disperse? Why is it trapped in very cold air masses?

Figure 5.9 (insert) makes clear that there is a substantial postwar warming during the cold portion of the year in Siberia and northwestern North America (and much less elsewhere). At various levels above the surface—from 5,000 feet all the way to the stratosphere—there is no warming at all in the last two decades. This is where and when we should expect the greenhouse effect to be rapidly toasting everything.

A comparison of winter half-year warming with the summer half-year in the Northern Hemisphere (Figure 5.10) shows a considerable

Figure 5.10
OVERALL NORTHERN HEMISPHERE POSTWAR WARMING, WINTER
AND SUMMER

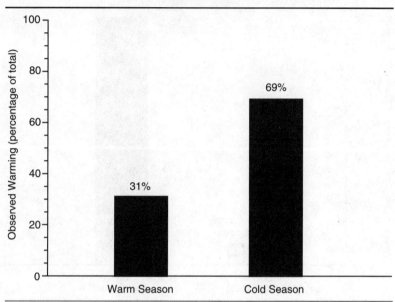

SOURCE: Michaels et al., 2000.
NOTE: The ratio of winter to summer warming is greater than two to one.

difference. Sixty-nine percent of the postwar warming is in the winter. How much of the winter warming was confined to the very cold high-pressure systems of Siberia and northwestern North America? The stunning answer? Seventy-eight percent. We just published these results in the journal *Climate Research* (Michaels et al., 2000).

Obviously, a warming of the very cold and deadly winter air masses is a pretty good thing; after all, winter temperatures are so far below freezing that a few degrees of warming could not possibly melt polar ice. A major summer warming would be more fearsome. How much of the warming of the last 50 years is in this exceedingly cold air? Do the math: 69 percent of total warming is in the winter; 78 percent of that warming is in the deadliest air masses (Figure 5.11). That means more than half of the warming is occurring in these air masses. They only cover, on a seasonally adjusted basis, 12 percent of the area. So warming is compressed—by a factor of

Figure 5.11
NORTHERN HEMISPHERE OCTOBER–MARCH WARMING

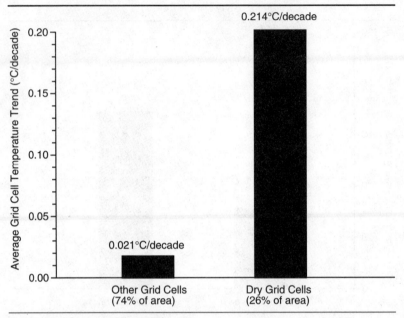

SOURCE: Michaels et al., 2000.

NOTE: The average winter half-year warming in cold dry air masses is more than 10 times the warming over the rest of the hemisphere.

four—into the most obnoxious air masses we know of, mitigating their deadly frigidity.

The air masses that have absorbed the lion's share of the warming average only around 4,000 feet in depth. As balloon (and satellite) records show, no warming whatsoever has occurred from 5,000 feet all the way to the top of the troposphere in the last 20 years, after allowing for the now-departed 1998 El Niño.

Climate models in the 1990 IPCC report predicted that the entire troposphere should be warming at the rate of about 0.4°C (0.75°F) per decade. The later sulfate-greenhouse versions, which serve as the basis for major emissions reductions proposals, dropped their warming to about 0.25°C (0.4°F) per decade by the addition of sulfate aerosol as a cooling factor, prompted by the embarrassing disparity that was developing between what was predicted and what happened.

Why is the warming confined to the bottom 10 percent of the troposphere? If the theory of transfer of infrared radiation is correct in the models, then 10 times as much warming should be observed if it is confined to one-tenth of the space. In other words, if warming is confined to the bottom 4,000 feet, we should be warming at about 2.5°C (4.5°F) per *decade* for sulfate-cooled models and 4°C (7°F) per decade for basic greenhouse models. Instead, we are seeing 0.15°C (0.27°C) per decade, largely crammed into a very small, very cold area in the dead of winter.

What is so bad about this type of warming? Can it be explained away by sulfate aerosols tempering greenhouse warming?

6. The Sulfate-Greenhouse Paradigm vs. the Reality of Climate Change

Let us stipulate the reigning paradigm in climate change: Greenhouse warming models predicted way too much heating, far too fast; but introducing a countering effect—cooling from sulfate aerosols—brings projections and observations much closer together.

On July 4, 1996, four days before a UN conference in Geneva at which Rio treaty signatories would agree to "binding targets and timetables" for greenhouse emission reductions, *Nature* made it official: The combination of greenhouse warming and sulfate cooling provided the definitive resolution to the climate change problem.

The article, by Lawrence Livermore National Laboratory scientist Benjamin Santer and 12 coauthors, was impressive and bold in its abstract:

> The observed spatial patterns of temperature change in the free atmosphere from 1963 to 1987 are similar to those predicted by state-of-the-art climate models incorporating various combinations of changes in carbon dioxide, anthropogenic sulphate aerosol and stratospheric ozone concentrations. The degree of pattern similarity between models and observations increases through this period. It is likely that this trend is partially due to human activities, although many uncertainties remain, particularly relating to estimates of natural variability.

Newspapers around the world picked up their finding, which was, said Australia's Neville Nicholls, a lead author of the 1996 IPCC report, the "clearest evidence yet that humans may have affected global climate."

The Santer article compared the temperature evolution in the tropospheric (5,000-to-40,000 foot) layer, which varies with latitude, and the stratospheric (above 40,000 feet) layer with the results of climate models that simply changed the greenhouse effect; with the results of models that changed the greenhouse effect and added the

93

cooling of sulfate aerosols; and the results of models that, in addition, depleted stratospheric ozone in a fashion that mimics what has happened as a result of (in part) chlorofluorocarbon refrigerants in the atmosphere.

The ozone effect really was not all that large and can be dismissed in a critical analysis of these results. The greenhouse-only models predicted a well-known signal, namely, amplification of warming in high latitudes and in the middle of the tropical troposphere, with pretty much equal effects between the Northern and Southern hemispheres. This model's results did not match the known reality that much of the Northern Hemisphere has been warming more rapidly than the Southern Hemisphere for much of the last 50 years.

When sulfates were added to the greenhouse mix, the model results were dramatically improved. Sulfates are produced primarily in the midlatitudes of the Northern Hemisphere as a byproduct of fossil fuel combustion. Unless they are specifically removed, most fossil fuels include some sulfur compounds that form sulfur dioxide when burned. This breaks down to sulfate, a finely divided particle that reflects away the sun's radiation and also helps to enhance cloudiness a bit. (Of the fossil fuels, natural gas produces little or no sulfate, and coal produces the most.)

Our figures (6.1, 6.2) show the remarkable correspondence between the sulfate-greenhouse model of Taylor and Penner (one of two used in this study) and the observed temperature changes Santer used, which run from 1963 through 1987. As chapter 4 revealed, this model uses only half the known changes in the greenhouse effect, and it is obviously way too warm when the real value is input.

The method Santer used is called pattern correlation, a mathematical construct that compares predicted departures from the global average at a given point with observed departures. Pattern correlation looks only at *relative* patterns. The pattern correlation statistic reaches high values in a sulfate-greenhouse model as long as the Northern Hemisphere midlatitude regions (where most of the sulfate effect is) cool compared with the midlatitude regions in the Southern Hemisphere. It also reaches high values as long as the stratosphere cools compared with the troposphere.

That is to say, the statistic would be high even if global *cooling* were occurring, provided that the mid-Northern Hemisphere cooled

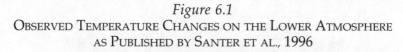

Figure 6.1
OBSERVED TEMPERATURE CHANGES ON THE LOWER ATMOSPHERE
AS PUBLISHED BY SANTER ET AL., 1996

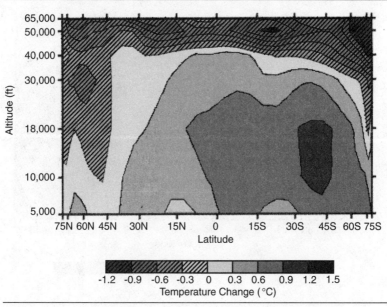

SOURCE: Santer et al., 1996.

NOTE: The warming in the Southern Hemisphere is much more rapid, especially between 5,000 and 30,000 feet.

more than the Southern and the stratosphere cooled more than the troposphere! The IPCC continues to use this and related methodologies today as it searches to confirm the "human signal."

The appearance of Santer's paper immediately before the Geneva conference prompted an extreme amount of skepticism on our part. Assuming its timing was not an accident, was there also something suspicious about its information?

More than anything else, the bulk of the pattern correlation between the model and reality has to do with the known strong cooling of the stratosphere that has been going on since measurements first began. While the current paradigm, expressed by K. Ya. Vinnikov in *Geophysical Research Letters* in 1996, holds that this cooling is simply a result of ozone depletion caused by CFCs, an inspection of the data shows that assumption to be false.

95

Figure 6.2
TEMPERATURE CHANGES PROJECTED BY THE TAYLOR AND PENNER
MODEL WITH SULFATE AND GREENHOUSE GASES

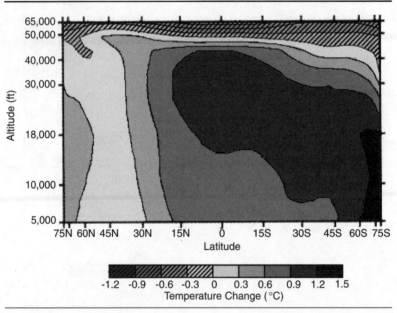

SOURCE: Santer et al., 1996.

NOTE: The correspondence with Figure 6.1 is nothing short of phenomenal.

Figure 5.6 showed lower stratospheric temperatures from the beginning of the record in 1958 through 1997. Scientists first observed the "ozone hole" and stratospheric depletions in the early 1980s. If cooling in this region is caused by ozone depletion, then the temperature decrease should begin around 1980. But when we look at the complete record, we see that it has been cooling since the beginning, although there are some very cold years at the end. This fnding is much more consistent with a greenhouse change going on for 40 years, with the addition of an ozone component in the last 15 years. (Note also the sharp warm spike from the 1997–98 El Niño.)

We examined Santer et al. to see how much of the correlation resulted from stratospheric cooling. Mathematically, when we look at the troposphere and stratosphere concurrently, the correspondence between the model and reality is a healthy 64 percent. But when we look at the lower troposphere, which is where we live, the correspondence drops to 16 percent. In other words, three quarters— (64–16) / 64—of the positive match between the model and reality is driven by changes miles above our heads.

Our plot of observed temperature changes between 1963 and 1987 (which is what Santer used; see his abstract) shows that in the troposphere, the largest warm anomaly—the one that contributes by far the most to the positive correlation between the model and reality beneath the stratosphere—occurs in the Southern Hemisphere between 30° and 60° latitude, from 5,000 to 30,000 feet.

We were perplexed that neither satellites nor weather balloons (our only two ways of measuring the tropospheric temperature) find any warming in the Southern Hemisphere during the last two decades. How did Santer et al. manage to see the big anomaly that is so obvious in their illustration?

The key lies in selecting the data, namely that from 1963 through 1987. The weather balloon record runs virtually through "yesterday," which, for Santer's 1996 paper, would mean through 1995. The data begin in 1958. One of the Santer paper's 13 coauthors, A. H. Oort, is an expert on this record, and his pre-1996 publications carry it into the early 1990s.

At the beginning of Santer's study in 1963, the Indonesian volcano Mt. Agung blew its stack, which was the largest explosion on the planet since Alaska's Katmai went off in 1912. Writing in the *Journal of Climate*, Angell and Korshover attributed to it a lower tropospheric cooling of 0.3°C to 0.4°C (0.5° to 0.7°). Volcanic coolings usually maximize in the year after the eruption, and it is no surprise that 1964, the second year in the Santer et al. record, shows an impressive downward spike compared with the rest of the data. In other words, they started at what should have been the coolest point in the record, rather than the beginning.

Figure 6.3 shows temperatures in that warm zone beginning in 1958 and ending in 1995. The warming trend, so evident in the period used by Santer et al., completely disappears when all the data are used.

We were surprised when *Nature* published our result, given its editorial stance on global warming. But we learned a bit about the political leanings of those who run that prestigious journal. *Nature* has a hard and fast policy that it will not publish anything that has

Figure 6.3
WARM REGION TEMPERATURE HISTORY, 1958–95

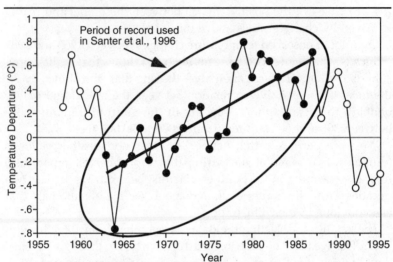

SOURCE: Michaels and Knappenberger, 1996.

NOTE: The closed circles represent the data used in the study, compared with the data excluded (open circles). The "human fingerprint" Santer et al. claimed in their study was largely a result of the years chosen.

appeared previously, either in another scientific journal or, more important, in the mass media.

Our analysis showing the "rest of the story" was completed on July 5, one day after the Santer story appeared. That is how easy it was to check! We were in a hurry because the Geneva Conference began on July 8. Michaels took a brief one-page write-up including the chart to Geneva and placed it on the literature table in the back of the conference. It was this document that got Under Secretary of State Timothy Wirth so mad when he spoke to the conference (of which more is in chapter 11).

By early September, our illustration had acquired some fame by appearing in *National Review*, the conservative newsweekly. We did not send it there, but we suspect someone at the Geneva meeting did. That *Nature* never pulled the paper is testimony that not one person there involved in editorial decisions pertaining to global warming submissions reads the conservative magazine!

When *Nature* published our piece in December 1996, Santer objected that the upper-air temperature record we used, from James Angell et al., was not the same as the one he used, from coauthor A. H. Oort. Yet a search of the literature produced a *Climate Dynamics* article in which Oort's record runs through 1993, not 1987. What is more, though there are a few years when the two records are not concurrent, the correlation between the years when the Oort and Angell records *are* concurrent is 0.94, meaning they are virtually the same data. Not surprising when you consider that the same weather balloons gathered much of the original data. Angell's record is slightly more sparse, but it reproduces regional temperature as well as Oort's—or, for that matter, the satellites; we published this in *Nature* also. Finally, Benjamin Santer himself is an author of a late 1996 article in *Science* that uses a complete upper air record, running virtually through publication time. A trace of the dates of submission and acceptance for Santer's *Nature* and *Science* articles proves that Santer was aware of the longer record by spring 1996; that record, too, behaves similarly to the Angell et al. history.

The satellite temperatures offer further insight into the sulfate aerosol question. If sulfates are responsible for an overestimation of global temperature, they must inordinately depress Northern Hemisphere temperatures. After all, there are very few people and consequently very few sulfates in the atmosphere of the southern half of the planet.

So greenhouse effect warming should be operating unfettered in the Southern Hemisphere, and temperatures there should be increasing with respect to the northern half of the planet.

But the satellite shows exactly the opposite. Figure 6.4 shows temperature departures from normal in the Southern Hemisphere subtracted from those in the Northern Hemisphere:

$$X = \Delta NH - \Delta SH \text{ where } \Delta = \text{departure from normal.}$$

If the Southern Hemisphere is warming more rapidly than the Northern, as would be the case for a sulfate-greenhouse scenario, then the departure from normal should be becoming increasingly negative.

Obviously it is not. In fact it is changing significantly in the *opposite* direction.

History Repeats Itself

In February 2000, Santer et al. appeared again in *Science*, this time saying that there is in fact no disparity between climate models and

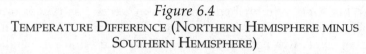

Figure 6.4
TEMPERATURE DIFFERENCE (NORTHERN HEMISPHERE MINUS
SOUTHERN HEMISPHERE)

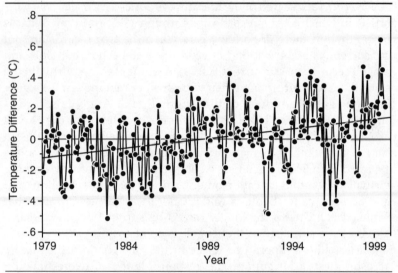

NOTE: The sulfate theory predicts that the amount of temperature difference should be going down as the Southern Hemisphere warms with respect to the Northern because of the Northern's cooling sulfates. But the opposite is occurring.

the satellite and balloon temperatures over the last 2 + decades. He made his argument largely by factoring in the effects of the big 1992 Mt. Pinatubo volcano and ending his study in 1998—right at the end of the huge warm spike caused by the giant El Niño.

Our experience with his 1996 paper made us immediately suspicious. First, there was another big volcano that occurred during the period in question, the 1982 eruption of El Chichon. When that is put in, the disparity between the models and the observations becomes larger. We can think of absolutely no good reason not to include both volcanoes, just as we could think of no good reason for only looking at 1963–88 data in the earlier *Nature* paper.

Further, the end point of 1998 is clearly an anomaly caused by El Niño. None of the models has this feature at that point in time. So it is better to adjust the record for not only the 1998 El Niño, but earlier ones, as Christy and McNider did with the satellite data in

Nature in 1994. When that is done, the disparity between his model and the observations reaches 0.15°C (0.27°F) per decade, which is very close to the observed difference between them. Inadvertently, by not using all of the data in a consistent fashion (i.e., selecting one volcano and ignoring another one), Santer again strengthened the scientific argument against dramatic warming.

Nothing New

In 1992, Michaels and his coworkers decided to ask what seemed, at the time, to be a logical question: Are sulfates really responsible for the overpredictions of global warming?

Then, the state-of-the-art climate model was the coupled ocean-atmosphere general circulation model (GCM) from Princeton's Geophysical Fluid Dynamics Laboratory (GFDL), a model that did not have sulfates in it. So Michaels decided to compare the GFDL's performance in both sulfate-free and sulfate-prone regions, asking whether the model tended to make its largest errors where one would expect sulfate cooling to compromise greenhouse warming.

Michaels was dismayed when a GFDL scientist told him that the lab would not supply the year-by-year results for the latitude/longitude grid. When he tried to get analogous output from the U.S. National Center for Atmospheric Research (NCAR) model, he was told the same. So much for open scientific inquiry. (This also led to our suspicion that there was something they did not want him to see.) Fortunately, a "mutual friend" had no problem when he asked GFDL for the data and he gave it to Michaels. (The NCAR data was not forthcoming at the time.)

Michaels found several "surprises" and wrote them up as an invited paper for the American Meteorological Society's 1996 annual meeting in Dallas, Texas, which he delivered before a standing-room-only crowd. The study broke down observed temperatures into their characteristic seasonal patterns and revealed that the GFDL model indeed captured the behavior of the most likely pattern of summer temperature change in the 20th century very well; however, that was the only one that was correct, with one exception.

He also broke down the data into the "sulfate" regions (eastern North America, Europe, and portions of Asia) and compared model performance with the more "sulfate-free" regions. The model's absolute best performance, bar none, occurred in the sulfate zones. Only one problem: The GFDL model did not have any sulfates in it!

101

In other words, the model was performing best in areas where it was supposed to do worst, assuming sulfates were the reason the models were overpredicting overall temperature changes.

Recall from chapter 4 that there are no real sulfates in a computer model; instead, the amount of the sun's radiation that goes to warming the surface of the earth is reduced by an amount that sulfates are supposed to intercept or reflect. The fact of the matter is that *any* computer command that reduces this amount of radiation will cool the planet (a core belief in climatology being that the sun warms the earth). Reducing the temperature of the Northern Hemisphere via this mechanism (for the Northern Hemisphere has most of the sulfates) is a convenient fix that may in the end signify nothing.

In a 1997 *Science* article on oceanic temperatures, M. A. Cane expressed the same notion concerning sulfates:

> It is quite possible that the influence of the values currently used in simulations is considerably larger than the true influence and is substituting for natural moderating mechanisms that are absent or underrepresented in present models.

Cane then cited NASA climate modeler James Hansen, among others:

> J. T. Kiehl and B. P. Briegleb [in *Science*] . . . and J. Hansen, M. Sato and R. Ruedy [in *Atmospheric Research*] demonstrated that the influence of sulfate aerosols is far too small to correct [the errors in] the [climate model] simulations.

One of the best ways to adjudicate a scientific disagreement is to go out and measure whatever is in dispute. That is precisely what the University of Washington's Peter Hobbs did when he measured the properties of sulfate-containing aerosols. Hobbs found that sulfate aerosols exist in an environment that also contains black soot, which warms things, and that the net effect is very close to neutral.

Sulfate Vacillations

More than any single scientist, NASA's James Hansen is credited with lighting the bonfire of the greenhouse vanities, with his June 23, 1988, congressional testimony that there was a "cause and effect" relationship between "the current climate and human alteration of the atmosphere." His testimony took place on the first day of summer in the middle of a scorching drought over much of the eastern

United States. In some places, it was the most severe dryness since the dust bowl of the 1930s.

Later that summer, Hansen published the results of his climate model in the *Journal of Geophysics*. His model was a bit warmer than most others and relatively coarse in resolution, but it really was not much different from the other GCMs coming on line at that time. And like the others, Hansen's model was soon recognized to be predicting too much warming, too fast.

So, two years later, Hansen published an update of the model in *Nature* in which he demonstrated that sulfate aerosols, in addition to greenhouse warming, provide a better explanation of recent temperature history.

In 1997, Hansen published a new modeling study in *Journal of Geophysical Research*, in which he said the "mean effect of [primarily sulfate] aerosols on surface temperature is nearly neutral . . . and their overall effect is less of a negative forcing [cooling] than has commonly been assumed."

Hansen called his new model "Wonderland," for reasons he has yet to explain. In this version, sulfates had a "direct" cooling effect (because they reflect solar radiation), an "indirect" cooling effect (because they enhance cloudiness), and a "semidirect" warming effect (based upon their physical appearance). Since no one has accurately measured any of these characteristics on a global scale (!), how they tweak a climate model becomes pretty subjective.

Less than one year later, Hansen published another paper, this time in the *Proceedings of the National Academy of Sciences*, that argued that his original (1988) model—which contained no sulfates—was right after all!

In the *PNAS* paper, Hansen was pleased to note that warming was occurring at about half the rate he had predicted in his previous "business-as-usual" scenario, published in 1988. The reason it warmed so little, Hansen says, is that the greenhouse gas emissions into the atmosphere were effectively changing the carbon dioxide at a much lower rate than previously expected, and the 1991 Pinatubo volcano was a cooling whopper. While most GCMs, Hansen's included, assumed "business as usual" meant an effective carbon dioxide increase of 1 percent per year, it turns out that the actual rate of increase is closer to 0.4 percent for the last 15 years (the "conservative" figure we use in this book is 0.7 percent). Why the slower rise? One reason is that the earth is getting greener as the increasing carbon dioxide stimulates plant growth. "Apparently," Hansen concurred, "the rate of uptake by CO_2 sinks, either the ocean, or more likely forests and soils, has increased."

Another reason warming is slower than projected is the stabilization of atmospheric methane Dlugokencky identified. Finally, Myrhe noted in 1998 that the direct warming effect from carbon dioxide itself was overestimated by 15 percent, as he reported in *Geophysical Research Letters*.

The evolution and devolution of Hansen's views shows the difficulties that ensue in trying to match the observed patterns of climate change to computer simulations. It looks as though the explanations for climate change are changing faster than the climate.

O, Inconstant Sun!

One of the long-discarded paradigms of climatology is that the sun is a constant star; we now know that it is slightly variable, with its output reaching the top of the atmosphere varying between 1,364 watts per square meter in the 17th century to 1,368 watts per square meter today, as Judy Lean reported in the *Journal of Climate*. This four-watt change is not inconsiderable but must be adjusted downward for three reasons: the earth's reflectivity; the differential absorption of radiation away from the surface (for example, the interception of the ultraviolet portion of the sun's output by ozone, mainly in the stratosphere); and the fact that half the planet is in shadow at any time. Lean calculates that the sun has influenced temperature on the order of about 0.5°C (0.9°F). In the last three decades alone, the change has been about +0.15 watt, or +.05 watt per decade, which translates to +0.02°C (0.04°F) per decade. When looking at observed temperature trends, we must subtract this amount to calculate the "human" component.

A Simple Explanation

In chapter 4, we noted the 1996 IPCC statement that, indeed, most climate models with greenhouse-only changes predicted too much warming too fast unless they assumed a "lower sensitivity" of the mean temperature to greenhouse changes. Then, of course, the IPCC conflated the observed warming deficit with sulfate aerosols.

Having shown that sulfate aerosols are not a sufficient explanation, we must default to only one conclusion: The models overestimated the sensitivity, and the planet is not going to warm that much.

You may wish to refer to Figure 4.2, which shows the tendency for all general circulation models to predict a linear or nearly linear warming for the next century, even though greenhouse gas input to the models grows exponentially. Call it the law of climate models: Despite exponential greenhouse forcing, *once greenhouse warming starts, it takes place at a constant rate.*

Another View?

While this book is being printed, a new paper, by Tom Karl, will appear in *Geophysical Research Letters* that will argue that the warmth of 1997 and 1998 requires that forecast warming for the next century be upped to 3.0°C (5.4°F).

At first glance this is doubtless the scariest global warming pronouncement made by a scientist of Karl's stature. But it has generated a chorus of negative reviews from other luminaries, including NASA's James Hansen, whose current position is very close to that expressed in this book; once greenhouse warming starts, it takes place at a constant rate.

Karl looked at the 16 consecutive warm months that dominated 1997–98, and argued that the projection of 2.0°C (3.6°F) of warming in the next 100 years is too low. This figure is currently the "central tendency" of the dozens of climate models that the IPCC will use in its next comprehensive summary, due out in 2001. Karl searched the model statistics and found that there is only a 5% chance that such a 16-month period would occur if the warming were going to be this low, but the chance goes up to 50% if the models forecast an additional degree (C) of warming in this century.

The period Karl studied coincides exactly with the huge 1997–98 El Niño. In fact, El Niño caused those warm temperatures. By the time Karl's paper was revealed, the globally averaged temperature, measured both by satellite and in NASA's surface record, had sunk to *below* its long-term average.

(continued next page)

(continued)

Karl went so far as to call 1997–98 a "change point" which shifts the rate of warming upwards, and he uses this to argue that the model projections must be similarly racheted higher. But, as shown in this book, the models don't have "change points"! So what Karl did was to add in a behavior that is simply not forecast. Perhaps he is right, and the mathematical form of every climate model is wrong. But, it seems, that basing a forecast based upon one very large El Niño is highly selective, especially as archaeological evidence argues that El Niños become weaker or less frequent with warmer temperatures.

That tenet is obvious both from theoretical points of view and based on practical observations. When dinosaurs roamed the earth, the carbon dioxide concentration was around 10 times higher than it was before the industrial revolution. And yet the temperature was only about 10°C (18°F) above what it is today. If the relationship between carbon dioxide and temperature were constant, then temperatures during that era should have been more than twice as warm. Not only that, but there is a lag between when the greenhouse effect changes and when the earth's oceans—which are very slow to warm—catch up.

When modelers factor in concepts such as these, GCMs produce a warming that is pretty much a straight line of constant slope, despite exponential increases in carbon dioxide. Recall from Figure 4.2 that the model rates vary greatly. This collection, which is as representative as any we can find, gives a mean warming of 0.25°C (0.45°F) per decade once warming starts. It does not matter whether it is the 1990s or the 2090s. *A straight line is a straight line.* Statistically, the two-thirds confidence limit about the trends is from 0.18°C (0.36°F) per decade to 0.32°C (0.58°F) per decade.

Figure 6.5 (insert) superimposes the IPCC's "global" surface temperature history on the model plots. (Remember that this is radically different from the history for the troposphere as a whole.) A close inspection of the record shows that warming starts around 1969. Since then, it has been pretty much a straight line, falling right around +0.15°C (0.27°F) per decade. But remember that the sun has been responsible for about .02°C of that per decade, which gives us 0.13°C (0.23°F) per decade left over.

For the moment, let us accept the IPCC's statement that "the balance of evidence suggests a discernible human influence on global climate"—by which they meant that human greenhouse (and sulfate) changes in the atmosphere are now affecting climate in a measurable way.

That the warming of the last third of the 20th century looks an awful lot like a straight line, coupled with the models' finding that, once warming starts, it maintains a constant slope, gives us the forecast for the 21st century. If we continue to emit greenhouse gases at the exponentially increasing rates that characterize the last third of the 20th century, we will warm up 10 times the recent decadal warmings, after allowing for the solar change (assuming solar constancy!), or 1.3°C (2.3°F). This is precisely the middle of the range Michaels presented in his first congressional testimony 10 years ago.

Ditto for winter vs. summer. In the last third of the 20th century, winter half-year (October–March in the Northern Hemisphere) warming has been about 40 percent more than summer. Both are also straight lines since 1965. So the expectation is that the 21st century will see a 1.45°C (2.61°F) rise in October–March temperatures and a rise of 1.15°C (2.07°F) in the summer half-year (April–September in the Northern Hemisphere).

These numbers change a bit depending upon what you believe. If you believe the sun will continue to warm as it did last century, results must be adjusted upward. Whatever occurs, we cannot control the sun's output, so our table allows for two options. The left-hand column assumes that Lean's calculations are right and that some solar warming must be subtracted from the observed 1965–97 trend; it also assumes the sun stays constant throughout the 21st century. The right column, on the other hand, warms the sun at the same rate it has warmed in the last 100 years. (We do not consider sulfates in this argument because they are so likely to have been overemphasized.)

Table 6.1
PLANETARY WARMING IN THE NEXT 100 YEARS: TWO VIEWS

	No future solar change	Sun continues to warm
Annual	1.3	1.5
Winter	1.5	1.7
Summer	1.2	1.3

NOTE: Warming (°C) to 2100 is based upon the models' linearity. "No solar change" means the sun's output remains constant in the 21st century but has warmed the atmosphere beyond human greenhouse changes in the last three decades. "Sun continues to warm" continues solar changes characteristic of the last three decades.

These numbers represent a convenient compromise between the IPCC and the "skeptics." In 1990, the IPCC estimated warming in the range of 1.5°C to 4.5°C; our figure, derived from nature, sits right at the bottom of this range.

Figure 4.1
PREDICTED TEMPERATURE CHANGES (°C), 1795–2040, RESULTING
FROM ALTERATIONS TO THE GREENHOUSE EFFECT

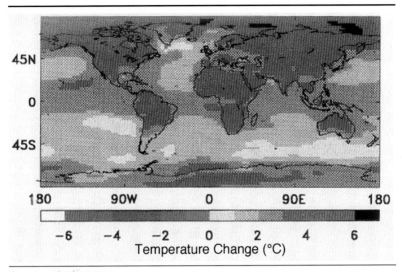

SOURCE: IPCC, 1995.

NOTE: The warming increases with latitude, especially in the Northern Hemisphere.

Figure 4.2
SIMILAR RESULTS: SEVEN TYPICAL GCM OUTPUTS DIFFERENT ASSUMPTIONS

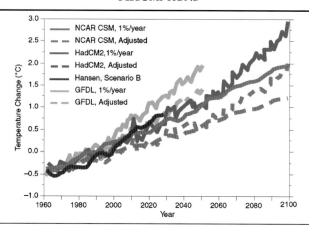

NOTE: "Hansen Scenario B" has the midrange of exponential greenhouse effect changes. "1%/year" means a 1 percent change in the carbon dioxide greenhouse effect that GCMs commonly use. But 0.7 percent is the more realistic figure that we have used as an adjustment. The point is that all the results look pretty much like straight lines, regardless of the assumptions made about the rate of change of greenhouse gases.

Figure 4.3
PREDICTED TEMPERATURE CHANGES, 1795–2040, WITH SULFATES FACTORED IN

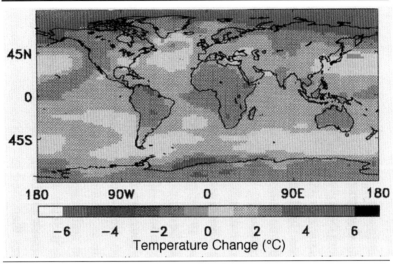

SOURCE: 1996 IPCC report.

NOTE: This is the same as Figure 4.1, except with sulfate cooling added in to greenhouse warming. The result is to drop the projected temperature rise in 2040 by roughly one degree C.

Figure 5.9
POSTWAR SURFACE WARMING IN THE COLD HALF-YEAR

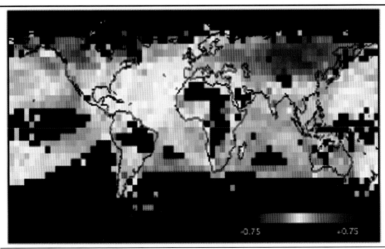

SOURCE: Balling, Michaels, and Knappenberger.

NOTE: Shows the concentration of warming in Siberia and northwestern North America during the cold half-year (October–March in the Northern Hemisphere, April–September in the Southern Hemisphere).

Figure 6.5
OBSERVED GLOBAL WARMING OF THE LAST THREE DECADES SUPERIMPOSED ON THE MODELS DEPICTED IN FIGURE 4.2

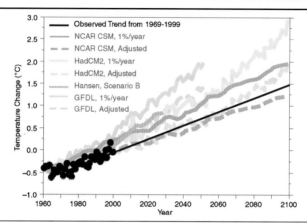

NOTE: The observed linear trend is near the lowest value that the climate models predicted and considerably below the mean predicted warming.

7. Greenhouse Projections of Circulation-Scale Changes vs. Reality

The title of a 1998 report from the prestigious climate modeling group at Germany's Max-Planck-Institut says it all: "Why Did the Greenhouse Warming Fail to Appear as Fast as Predicted?"

The MPI study fingers the usual suspects: sulfate aerosol, changes in the sun's activity, ozone depletion, and so forth. But the fact is, researchers still cannot determine what caused the lack of warming. So, the MPI scientists concluded that something must be wrong with the *observations*.

The disconnect between the greenhouse-only models and reality was known to be large even when the Rio climate treaty was signed in 1992. Based on original research he conducted in 1992, at the time of the Earth Summit, Michaels published the disparity between Northern Hemisphere temperatures and the GFDL model (at that time, the most prestigious of the GCMs) in a 1993 *Bulletin of the American Meteorological Society*, from which Figure 7.1 is reproduced here.

In 1995, British modeler John Mitchell, who was in charge of the United Kingdom Meteorological Office model (now called the Hadley Center Model), published a paper that implied that if sulfates did not cool the temperature, global readings should have risen 1.8°C (3.6°F) by now. But the predominance of land in the Northern Hemisphere (the surface of the Southern Hemisphere is 90 percent water) results in a more rapid warming, so that it is reasonable to assume that the Northern and Southern hemispheres should have warmed about 2.3°C (4.1°F) and 1.3°C (2.3°F), respectively.

The disconnect between the greenhouse-only models and reality was obvious as early as 1983 to any scientist who had carefully inspected the data, and the weakness of the subsequent sulfate-greenhouse "fix" was discussed in the previous chapter. Here, we explore different aspects of the climate, including features of the

Figure 7.1
OBSERVED VS. PREDICTED WARMING IN THE GFDL (NONSULFATE)
MODEL

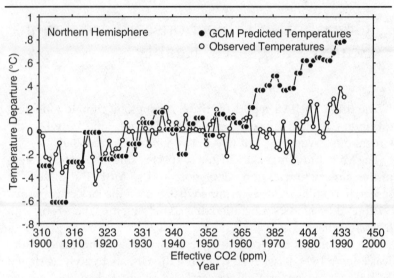

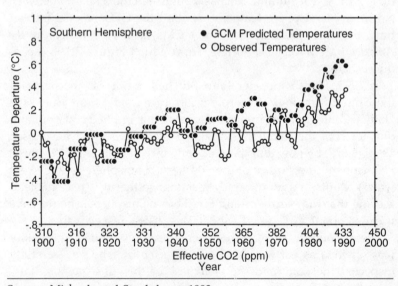

SOURCE: Michaels and Stooksbury, 1993.

atmosphere's general circulation and ask if things have changed in a fashion consistent with an anthropogenerated warming.

Climate's Increasing Variability

Those who promote a negative vision of climate change will, when pressed, usually admit that the planet did not warm as originally predicted. But they often cite the IPCC's statements about "more severe droughts and/or floods" and a "discernible human impact on global climate" as indicators that weather has become more variable and people are to blame. As a result, a slew of recent reports (the most recent, as of this writing, being *NBC Nightly News* on January 4, 2000) would have us believe that the increase in the concentration of atmospheric carbon dioxide will produce an increase in the overall variability of climate, thus exposing us to more climate extremes.

Though it is true that a few modeling studies suggest that an increase in variability is possible given a buildup of CO_2, others suggest a decrease. More important, the observed evidence generally shows a decrease in variability.

Gregory and Mitchell published one of the most telling papers on this subject in 1995 in the *Quarterly Journal of the Royal Meteorological Society*. They concluded that changes in variability could differ greatly from season to season and were highly dependent upon local physical processes. Under doubled carbon dioxide conditions, their numerical experiments revealed decreases in temperature variability in Europe in winter as a result of reduced land–sea thermal contrast, but variability increased in summer because of surface heat balance. Complex temperature variability results, like those found in Europe, could be expected in other parts of the world. Basically, it is difficult to look at the various model results on this issue and conclude that temperature variability will increase in the future.

When models produce mutually inconsistent projections, it is instructive to look at real data. In 1992, David Parker and two coworkers examined the world's longest thermometer record, known as the Central England History. (Just how long is it? It begins in 1659!) Parker found no evidence of increased variance in recent decades. In a different study, published in 1994 in the *Journal of Geophysical Research*, Parker also examined global records, such as

111

the report the IPCC published comparing interannual seasonal temperatures from the 1954–73 period with the 1974–93 period over most of the globe. He found a small increase in variability overall, with an especially large increase in central North America.

Tom Karl, now director of the U.S. National Climatic Data Center, wrote in *Nature* in 1995 that an increase in CO_2 should decrease temperature variability; indeed he and colleagues R. W. Knight and N. Plummer found that day-to-day variability during the 20th century is down in the Northern Hemisphere, particularly in the United States and China. In another paper published in 1997 in *Scientific American*, Karl stated that "projections of the day-to-day changes in temperature are less certain than those of the mean, but observations have suggested that this variability in much of the Northern Hemisphere's midlatitudes has decreased as the climate has become warmer. Some computer models also project decreases in variability."

A review of the literature makes it difficult to believe that scientists are predicting an increase in variability for a rise in CO_2. In a 1998 article in *Climate Research*, we analyzed trends in the variability of daily and monthly near-surface air temperatures in the 5° latitude by 5° longitude IPCC temperature record. We found that the trend in temperature variability within each year is toward reduced variability and is highly significant over the past 50 and 100 years (Figure 7.1). Furthermore, we found that temperature variability and global temperature are negatively related (Figure 7.2), with a tendency for lower variance in warmer years.

We then examined daily maximum and minimum temperatures from the United States, China, and the former Soviet Union for day-to-day variability in January and July. Most of the trends also indicated declining variability (Table 7.1). A final approach examined the occurrence of record-setting daily maximum and minimum temperatures from the same countries. We found no evidence for an increase in record temperatures.

In the journal *Physical Geography*, Balling (1998) examined changes in the spatial variability of mean monthly and daily temperatures that have occurred during the period of historical record. This study looked at satellite-measured temperatures as well as the more common surface records. The satellite temperatures, which are characteristic of the 5,000–30,000-foot layer, begin in 1979. The surface records were divided into two subsets, 1947–96 and 1897–1996. The study

Figure 7.2
Intra-annual Temperature Variability, over 100 Years

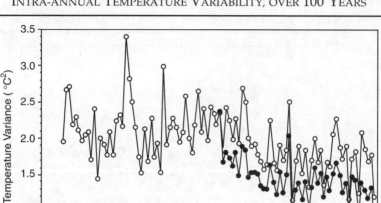

SOURCE: Michaels et al., 1998.

NOTE: This figure incorporates 393 valid 5° latitude by 5° longitude cells beginning in 1897 (open circles) and 1,041 cells beginning in the 1947 time series (filled circles).

was confined to the Northern Hemisphere because of the relative lack of surface data (a problem the satellite does not have) over much of the southern half of the planet. These subsets show relatively consistent temporal patterns with respect to geographic variability.

Overall, the geographic variability in temperature anomalies has declined during the period of historical records, and the interannual variability levels in temperature anomalies are negatively related to mean hemispheric temperature. The less things vary from place to place, the less they vary from month to month. We achieved this result by combining two separate studies we published in back-to-back issues of *Climate Research* (Balling, Michaels, and Knappenberger 1998; Michaels et al., 1998). We found that as the planet warms, the month-to-month swings also decline.

Clearly, there is little support for the popular perception that our temperatures have become more variable. And indeed, the preponderance of the evidence points to lower variability in temperature, both over time and across space.

Figure 7.3
INTRA-ANNUAL TEMPERATURE VARIABILITY VS. GLOBAL
TEMPERATURE ANOMALIES FOR THE 1897–1997 TIME SERIES

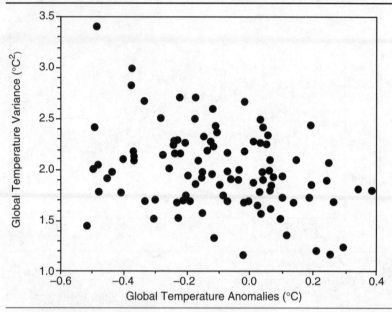

SOURCE: Michaels et al., 1998.
NOTE: The warmer the surface temperature, the less variable things become.

Table 7.1
DAY-TO-DAY TEMPERATURE VARIABILITY FOR THE UNITED STATES,
CHINA, AND THE SOVIET UNION, JANUARY AND JULY

Mean Linear Trend (°C/decade) in Daily Temperature Variability Values			
Temperature	USA	China	USSR
Maximum:			
January	− 0.19	− 1.13	0.07
July	− 0.13	0.06	− 0.02
Minimum:			
January	− 0.26	− 1.32	− 0.37
July	− 0.19	0.06	− 0.08

SOURCE: Michaels et al., 1998.

Figure 7.4
GEOGRAPHIC VS. INTERMONTHLY TEMPERATURE VARIABILITY

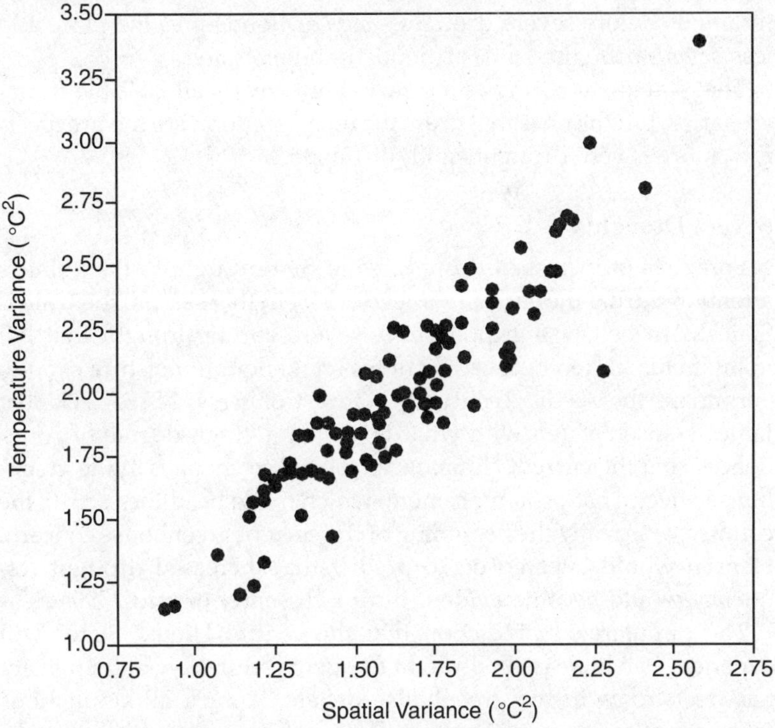

The result shown here along with that of the previous figure, indicates that warming reduces month-to-month changes as well as year-to-year.

Those who excerpt the IPCC's statements about increases in droughts "and/or" floods and the human influence are doing so quite selectively. For in the same report, the IPCC summarizes the debate in this way: "Overall, there is no evidence that extreme weather events, or climate variability, has increased in a global sense, through the 20th century, although data and analyses are poor and not comprehensive."

Precipitation Extremes
Changes in the patterns of precipitation across the globe have long been put forward as one of the major consequences of global

warming. In the 1996 IPCC report, the outlook for the future was summed up as follows: "Warmer temperatures will lead to a more vigorous hydrological cycle." That statement was followed by the infamous "more severe droughts and/or floods in some place and less severe droughts and/or floods in others" line.

That statement seems as if it just about covers all possible future scenarios, but the ones that draw the most attention are the prospects for "more severe droughts and/or floods."

Severe Droughts

Forecasts of increased drought were first brought to the public's attention during the early summer of 1988. In that season, the United States corn belt was gripped by a severe spring drought that left many fields barren because there was not enough moisture even to germinate the seeds. Testifying before Congress, NASA scientist James Hansen stated with what he called a "high degree of confidence" that the current climate was related to changes in the greenhouse effect. That pronouncement, which made headlines across the country, was really the beginning of the current greenhouse concern. Hansen would later go on to predict that increased drought frequency would become evident during the early or mid-1990s.

The popularized prediction that the central United States will experience an increase in drought frequency, duration, and intensity has deep roots in the greenhouse debate. Almost all simulations from numerical climate models show an increase in the drought risk in the heartland for the doubled carbon dioxide condition. The Intergovernmental Panel on Climate Change, in both 1990 and 1996, continued to warn that the midlands will dry in the future. Not surprisingly, officials and activists openly discussed many policies designed to deal with this threat to the nation's breadbasket. To this day, the numerical models, even when modified, continue to predict an increase in drought conditions through the 21st century.

Given that prediction, given the large size of the interior of North America, and given that droughts are a natural part of the region's climate system, frequent opportunities arise to show drought conditions to the public and reinforce the link to the greenhouse effect.

Long before his now-famous 1988 statement before the U.S. Congress, James Hansen in 1981 provided an overview in *Science* of the types of climate changes we might expect for increasing levels of

116

carbon dioxide. Among all the other changes discussed, Hansen mentioned that droughts could be increasing in North America. To bolster his computer model–based arguments, he noted that the period of about 5,000 years ago was a warm time, with planetary temperatures of 1°C to 2°C above those of today and dry conditions in North America.

In that same year, Manabe et al., running the GFDL climate model, reported on the results from three different climate models comparing the present climate with the simulated conditions for a quadrupling of atmospheric carbon dioxide levels. Manabe's relatively simple models for four times current CO_2 showed a substantial reduction in the soil moisture in the midlatitudes—up to 60 percent was lost in the summer season for the 40°–50°-north latitudinal band. Examination of the model output showed that this large decrease in soil moisture was related to less snow in the winter season, a rapid melting of the snow that did accumulate, and less rainfall in the summer season. When coupled with the 1981 Hansen article, a theoretical foundation for linking American droughts to the greenhouse effect emerged.

A half-decade later, but still before the popularization of global warming in 1988, Manabe and Wetherald, in 1986 and 1987, published two important papers in *Science* and *Journal of Atmospheric Science* in which they used two model configurations to evaluate the impact on soil moisture of both the doubled and quadrupled carbon dioxide. One of their model configurations had unchanging cloudiness while the other allowed a series of feedbacks to alter the clouds represented in the model. Their simulation experiments showed that the Great Plains of the United States experienced a 50 percent reduction in soil moisture for a doubling of the greenhouse gases when the planetary temperature rose 4°C.

Their variable cloud simulation showed that this decrease in soil moisture was due to a variety of factors including (a) the early termination of snowmelt; (b) an earlier ending to the normal rainy period in spring; (c) a decrease in cloud cover, allowing more sunlight to penetrate to the surface and thereby increasing potential evapotranspiration rates; and (d) a reduction in summer rainfall. Those two major papers provided support for rash claims that the 1988 drought was related to the buildup of greenhouse gases.

British modelers J. F. B. Mitchell and D. A. Warrilow published a 1987 paper in *Nature* entitled, "Summer Dryness in Northern Midlatitudes Due to Increased CO_2." Yet the bulk of their article dealt with the uncertainties in the predictions and did not provide much support for the connection between drought and the greenhouse gas increase.

Using the UKMO climate model, they found that increasing greenhouse gas concentrations led to decreasing soil moisture levels in midlatitudes in the Northern Hemisphere during the summer months but increasing levels of soil moisture in the winter. Mitchell and Warrilow concluded that the representation of the physical properties of the soils was critical to the response of soil moisture to increasing planetary temperature. Specifically, they showed that the numerical routines designed to handle runoff from frozen soils were quite important in determining expected trends in soil moisture levels. Despite the rather leading title of their article, the authors provide many reasons to question the validity of earlier results on drought.

In an attempt to anticipate changes associate with global warming, Kellogg and Zhao in the premier volume of the *Journal of Climate* (1988) used five different general circulation models in their evaluation of drought increases for a doubling of greenhouse gas concentrations. Three of the models they selected showed a decrease in soil moisture in the summer months in the central United States. The other two models had some peculiar features that may have prohibited the drying from taking place (e.g., poor land surface simulations or unrealistic dryness in spring). They compared the models for the winter season as well, and they found much more agreement in the winter simulations than the summer ones.

McCabe et al. (1990), in *Water Resources Bulletin*, took the temperature and precipitation output from three different doubled CO_2 model experiments and calculated what is known as "potential evapotranspiration." This rather complicated term tries to account for the amount of moisture that either evaporates from the surface or transpires through plants, given a constant supply of moisture.

All three models predicted a reduction in soil moisture throughout the United States. In an absolute sense, areas around the Great Lakes and New England showed the greatest decrease in moisture levels, but in relative terms, the northern plains had large decreases in soil moisture levels.

As the literature that linked the greenhouse effect to droughts in the United States was growing, an article appeared in *Climatic Change* in 1990 that was as scary as anything published in the professional literature to date. Hansen's employee, David Rind, along with several coworkers, conducted doubled CO_2 simulations with Hansen's model and focused attention on potential future droughts in the United States. Rind included routines in the model for calculating the future Palmer Drought Severity Index.

Rind's results were nothing short of frightening and certainly fed the media's collective memory when it confronted the 1999 weather in Washington, D.C., nine years later. He calculated that severe droughts would occur 50 percent of the time by the 2050s (an increase in drought frequency and intensity that should, Rind predicted, become noticeable in the climate record during the 1990s). His model simulations warmed the earth by 4°C for the doubled CO_2 experiment, and the increase in temperature drove up the rate of potential evapotranspiration leading to the large decline in soil moisture. Rind et al. stated that drought increases simulated in earlier experiments by other scientists were probably underestimates due to the lack of realistic land surface models. Finally, Rind et al. noted that evidence abounds in the paleoclimatic record for the connection between a warm earth and drought in the central portion of the United States.

Measuring Drought with the Palmer Index

The Palmer Index is the standard measure of drought. When people on the Weather Channel talk of "severe" or "extreme" drought, they are using the Palmer Index.

What exactly is a drought? *Merriam Webster's Collegiate Dictionary* says, "a period of dryness especially when prolonged." "Dryness" means "lacking or deficient in moisture." By that definition, most deserts are in a state of semipermanent drought and most forests are not. And that is probably true.

But that is not the way we climatologists measure drought. Instead, we employ the at least debatable idea that drought is measured primarily by the departure from average moisture at any spot. For example, if a place such as Washington, D.C., receives only 60 percent of its average rainfall, it is likely to be in a big-time drought, even though it still is getting 13 times the average rainfall of Death

119

Valley, which does not experience a Palmer-defined "drought" unless rainfall there is below its paltry 1.89-inch annual average.

Indeed, some of the rainiest places in the country, such as Washington's Olympic Peninsula, can receive scores of inches of rain per year, yet because this is substantially below normal, still be in a severe drought.

There is a certain sort of egalitarian logic around this notion. The implication is that the natural vegetation as well as the human infrastructure put in its place are adapted to "normal" conditions for that spot; the more abnormal things become, the more they fall apart. In other words, Death Valley Scotty is adapted to Death Valley and Al Gore is adapted to Washington, but if either experiences below-normal rainfall for his area he is unhappy.

Many different measures go into the Palmer Index. After all, water either precipitates (as rain), runs off (as a creek or a river), goes into the ground, or evaporates. Each of these processes determines the local moisture departure from normal.

Evaporation is exceedingly important in this equation. In the Mid-Atlantic region of the United States, a pan of water left out in December will evaporate about half an inch during the entire month, on average. In July, however, this figure goes up to seven inches, or 14 times as much. Evaporation is highly dependent upon temperature, which itself is highly dependent upon sunlight. Evaporation is the main variable driving some of the GCMs into summer drought; while many indicate rainfall unchanged from current conditions, the predicted temperature rises increase evaporation enough to create a drought.

Both latitude—which determines how high the sun is in the sky during summer—and temperature determine evaporation. All the evaporation variables are also linked in a "positive feedback loop." In general, the sunnier it is, the warmer it is, which promotes more evaporation, which makes it drier. The drier it is, the hotter it gets, as less and less of the sun's energy is required for evaporation; the hotter it gets, the drier it is; and so on.

The Palmer Index attempts to account for all these variables and then subjects them to the notion that departure from the average value is much more important than absolute dryness. As a result the Palmer Index is statistically "normalized." For any given location in the nation, the Palmer Index is assigned an average value of zero.

Readings that are statistically more than one "standard deviation" from the mean are given values of plus (for wetness) or minus (dryness) 2.0. On the dry side, all readings between −2 and −3 are assigned the moniker "moderate drought."

This has interesting consequences. Statistically speaking, one third of all the Palmer values are more than one standard deviation from average. For values below the mean, this means that at any given time, one sixth of the country, theoretically, is in moderate drought!

The statistics for "extreme drought" (Palmer value of more than two standard deviations below the average, or less than −4.0) work out similarly. Because the index is forced to a mean of zero and a constant standard deviation, on the average, 2.5 percent of the nation should be suffering "extreme drought" at any given time.

These theoretical averages have to be adjusted with reality, like most theories. It turns out that, in any given month, 4.4 percent of the United States shows a Palmer Index of less than −4.0.

This has tremendous consequences and subjects the Palmer Index to profound abuse, as our sidebar shows.

A Dearth of Drought

Consider the summer of 1999, when an intense drought hit portions of the Mid-Atlantic region. Newspapers carried it on the front page for weeks. All the major television networks led off with it for days on end. President Clinton, in a naked conflation of the drought with global warming and the need to "do something," said, on August 6, "As weather disruptions become even more common, and they will, they will demand a more coordinated response by the national government."

Figure 7.5 shows the area of the nation experiencing extreme drought (Palmer Index of −4 or lower) as Clinton spoke. On August 7, it was 1.98 percent of the lower 48 states, *or less than half the amount seen in a "normal" period.* Perhaps the drought's pretty much circumscribing itself around Washington, D.C., had something to do with the media hype of what in fact was perfectly unremarkable from a climatological standpoint. Perhaps the need to see every weather event as evidence for

(continued next page)

(continued)

global warming was involved too. Ironically, global temperatures measured both by satellites and weather balloons in the
summer of 1999 were below the average for the last 20 years.

Imagine the media squall if we had a recurrence of the
drought of 1934 (Figure 7.6). Nearly half the nation was suffering extreme drought or worse, and it was not even "caused
by" global warming!

Figure 7.5

PALMER INDEX MAP SHOWING 1.98 PERCENT OF 48 STATES IN
"EXTREME DROUGHT," AUGUST 7, 1999

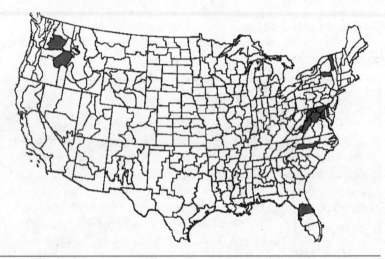

NOTE: The darkest regions are in "extreme drought," and comprise 1.98
percent of the nation. The average value is 4.4 percent. Despite the hype,
there was less drought than average at this time.

There are five reasons that the doubled CO_2 drying appeared
to be common in the model runs. Together they seem to make a
convincing picture.

1. Increased precipitation predicted for the winter falls on already
saturated soils, runs off, and is never available to the soil in the
summer season.

Figure 7.6
THE 1934 DROUGHT, AFFECTING NEARLY HALF THE SAME 48
STATES

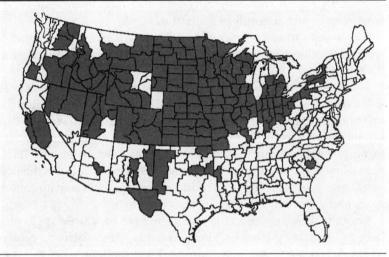

NOTE: In 1934, nearly half the nation was in extreme drought, or 25 *times* the area suffering extreme drought in August 1999 (Figure 7.5).

2. The continental interiors have a relatively large seasonal cycle in soil moisture in the present climate condition. In the doubled CO_2 experiments, the earlier start of the drying season accentuates the normal pattern, and lower soil moisture results.

3. Snow melts earlier in the doubled CO_2 simulations, and that moisture is lost more quickly and therefore not made available in the summer season.

4. Reduction in cloud cover increases potential evapotranspiration rates in summer, further lowering soil moisture levels.

5. An associated reduction in summer precipitation further exacerbates the soil moisture depletion in these continental interiors.

In 1992, Rind published another paper in *Nature*. To create a surrogate of greenhouse warming, he increased the planetary temperature by 4.5°C (8.1°F). Fifteen sites in North America showed a 35 percent decrease in soil moisture on an annual basis and a near 50 percent decrease in soil moisture in July. Rind recognized that these issues are complex and that more work is badly needed before much confidence could be placed on the results.

123

Wetherald and Manabe in the 1995 *Journal of Climate* conducted various quadrupled CO_2 computer experiments and concluded that summertime drying in the continental interior would be caused by local and regional increases in potential evapotranspiration over-whelming any changes in precipitation.

Fifteen years after the first major articles on this subject, the models were still calling for more droughts in the future in the central United States.

Again, by the time of the 1996 IPCC report, it had become fashion-able to include the sulfate aerosol cooling in the doubled CO_2 numeri-cal experiments, even though there are good reasons to suspect that sulfate aerosols really cannot explain the planet's reluctance to warm as rapidly and as much as initially forecast. Generally, the IPCC noted that the inclusion of sulfate aerosols lessened the reduction in soil moisture resulting from reduced warming, lower evaporation rates, and enhanced summer precipitation levels.

After the IPCC report, in the 1997 volume of the *Journal of Climate*, Gregory et al. conducted GCM experiments with a focus on central North America and southern Europe. Their model results still pre-dicted a decrease in both precipitation and soil moisture in these areas for the doubled CO_2 scenario. They found increases in winter and spring evaporation rates, less snow cover existing in the spring season, and temperatures increasing over these no longer snow-covered areas. The resulting decrease in local soil moisture levels in these areas leads to less rainfall in summer and a greater chance for drought.

Reality vs. Predictions of Drought

All of these scary predictions beg for a comparison with reality. If there is a substantial disconnection between the two, what does that say about all the model-based results?

As with so many other areas in the global warming debate, there is considerable evidence arguing against the prediction of increased drought in the central United States. Here we find a classic problem: voluminous literature predicting an increase in drought conditions as the greenhouse gas concentration increases; and yet, during the past century (and during a time of considerable greenhouse gas buildup), a trend away from drought and toward more moist condi-tions in the North American interior.

Figure 7.7
PERCENTAGE OF UNITED STATES IN SEVERE DROUGHT

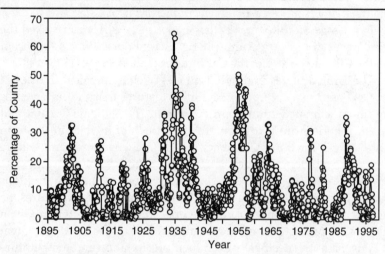

NOTE: The percentage of the United States experiencing severe or extreme drought conditions fluctuates from year to year but shows no long-term trend.

What the models characterized as an inevitability simply has not happened. Figure 7.7 shows the percentage of the United States that has experienced severe to extreme drought conditions each since January 1985. There is no long-term trend in this record.

Looking back much further in history, Laird et al., in 1996 in *Nature*, showed that drought frequency and intensity were much higher prior to 1200 A.D. in the north-central United States than they have been during the past century. The empirical evidence does not appear to support a dire prediction so consistently made by computer models.

In other parts of the world the results are similar. In some places, such as the African Sahel (the destitute zone immediately to the south of the Sahara Desert), there has been a tendency toward more dry conditions, but overall, according to the 1996 IPCC Report, there is little evidence for changes in drought frequency or intensity.

Recent Empirical Research

Since the prediction of drought in the central United States became part of the vision for the future, many climatologists have analyzed historical drought records in search of the greenhouse signal. Despite all of the computer predictions, *virtually every investigator has found no trend at all, or even a trend toward increasing moisture levels.*

In a *Geophysical Research Letters* paper in 1989, Hanson asked the titular question, "Are Atmospheric 'Greenhouse' Effects Apparent in the Climate Record of the Contiguous United States (1895–1987)?"

The United States is divided into 344 "climatological divisions" (CDs) based usually upon geographic features that should give rise to similar climate within a particular unit. The National Oceanic and Atmospheric Administration archives much of its data in the form of averages for those CDs, back to 1895. It was those temperature and precipitation records that Hanson examined, looking at data from 1895 to 1987 for all 344 climate divisions in the conterminous United States. After using a variety of identification procedures, he found no overall trends in either temperature or precipitation. Given the GCM predictions of increasing aridity in the northern Great Plains, Hanson specifically examined precipitation and temperature trends for that part of the country. He found no trends in either variable; further, he created an index of winter vs. summer precipitation, and again, he could identify no trends in the data. Nothing in his analyses would lead anyone to believe that droughts were becoming more frequent or more intense in the central part of the United States.

The following year, Karl and Heim, also in *Geophysical Research Letters*, published another article titled in the form of a question: "Are Droughts Becoming More Frequent or Severe in the United States?" They examined the Palmer Hydrological Drought Index, a variable very similar to the Palmer Drought Severity Index. Again, the data came from 344 climate divisions in the conterminous United States for the period 1895–1989. They found that drought conditions in the United States worsened from 1895 to the mid-1930s, but that since the mid-1930s, the PHDI values indicate a trend away from drought and toward increasing soil moisture levels. Karl and Heim concluded that there is "no evidence to support the notion that the frequency, area extent, or severity of droughts in the United States is increasing."

Peter Soulé published a series of four papers in the refereed literature in the early 1990s regarding trends in drought conditions observed over the past century in the United States. Soulé has used a variety of drought-related climate variables and composite indices, time periods, spatial units, and statistical procedures. He has found that trends upward or downward in drought records are highly dependent on the period of record and the area of interest: Choose the right starting and ending years and choose the right area in the United States, he discovered, and you can identify various statistically significant trends in the drought conditions. Overall, Soulé's work has been quite consistent with Karl and Heim and many others: The GCMs' projections notwithstanding, the expected trend toward increasing drought frequency and duration is not occurring in the United States.

The 1990 IPCC report only briefly addressed the issue of changes in drought conditions around the world. Examining data from 1890 to 1990, the IPCC concluded that "drought and moisture indices calculated for Australia, parts of the Soviet Union, India, the USA, and China do not show systematic long-term changes."

With Sherwood Idso, Balling examined precipitation records from 934 largely rural stations in the conterminous United States from 1901 to 1987 in a 1991 paper in the journal *Environmental Conservation*. They found little trend from 1901 to 1954, but from 1955 to 1987 the rainfall for all stations increased by 8.6 percent. In the central United States, the increase in precipitation from these rural stations was 16 percent during the most recent 33 years of record. Such an increase in rainfall is not likely to be associated with an increase in drought, unless summer temperature has increased dramatically (which it has not).

In another paper, published in 1992 in *Agricultural and Forest Meteorology*, we looked specifically at the drought patterns for the conterminous United States from 1895 to 1989 and found a statistically significant *increase* in moisture levels from 1955 to 1989. The change was greatest in the central United States with Colorado, Nebraska, Kansas, Oklahoma, Iowa, and Missouri having some of the largest increases in moisture levels. Once again, the evidence for the predicted drought signal, this time in the central United States—the region about which the IPCC was so concerned—was missing.

In two comprehensive reviews of U.S. climate trends, Tom Karl and his coworkers acknowledged that climate models are predicting more severe and longer-lasting droughts in the central United States, particularly during the warm season; these reviews were published

127

in 1995 and 1996 in *Consequences* and the *Bulletin of the American Meteorological Society*, respectively.

Then they examined precipitation records from thousands of stations across the United States and found that since 1970, precipitation has been about 5 percent higher than the mean level for the 70 previous years. Rainfall was generally increasing throughout the central United States, although some curious features appeared, such as large increases in rainfall in South Dakota but declining precipitation in North Dakota. Rather than calculate the trend in areally averaged Palmer Drought Severity Index (PDSI) values, Karl et al. determined the portion of the country experiencing severe drought or moisture surplus each year. Consistent with the precipitation data, they found that since 1970, the United States has shown a jump toward moisture surpluses.

Other data are available besides the usual temperature and precipitation figures. Tom Peterson, in a 1995 *Nature* paper, examined the evaporation measurements from 746 stations in the United States for the period 1945–90, and found a significant decrease. The trends were downward at most collection sites, an average of 4 percent. Peterson argued that the trends were associated with an increase in low clouds that was cutting down on incoming solar energy and the resultant evaporation rates. Note that low clouds provide a net *cooling* effect. While the models show that evaporation rates should be increasing with the buildup of greenhouse gases, actual measurements show a decrease during the period of observation.

With U.S. Geological Survey scientist Harry Lins, Michaels examined stream flow records in the United States in the last half of this century in a paper published in the *Transactions of the American Geophysical Union* in 1994. Everything else being equal, an increase in stream flow means that evaporation and drought must be decreasing. Wherever they found a significant change, it was in the positive direction. There was simply no evidence for increasing drought in the stream flow histories.

In 1996, Balling assembled the Palmer Drought Index values from 1895 to 1995 on a state-by-state basis. His study concluded, "there is little indication of any trend to increasing drought conditions" across the United States.

IPCC's Prediction Fails

The 1990 IPCC report specifically selected an area in the central United States for a series of detailed climate predictions. The area

Figure 7.8
20TH-CENTURY MIDWESTERN U.S. TEMPERATURE CHANGES

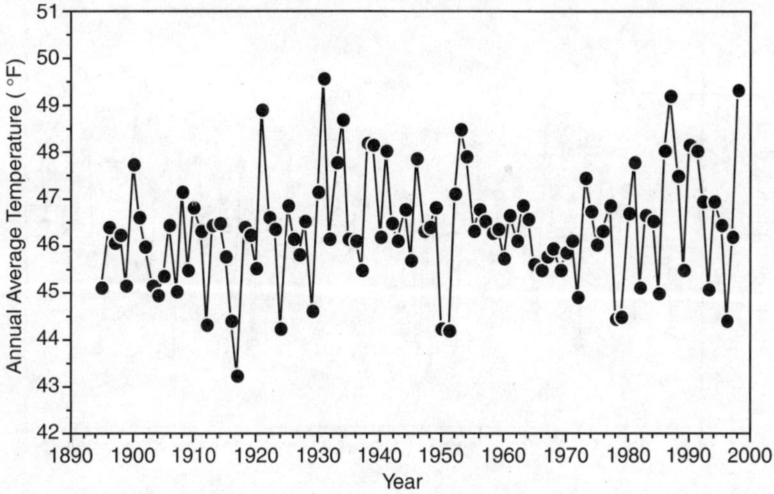

NOTE: No change is evident consistent with GCM projections.

selected was bounded by 35° north to 50° north latitude and 85° west to 105° west longitude—essentially central Oklahoma to southern Manitoba and the Ohio–Indiana border to eastern Wyoming. According to the IPCC document, the doubling of greenhouse gas concentrations will be associated with a warming of 2°C to 4°C (3.6°F to 7.2°F) in winter and 2°C to 3°C (3.6°F to 5.4°F) in summer. Precipitation increases range from 0 to 15 percent in the winter months whereas summer shows a 5 percent to 10 percent decrease in rainfall. Soil moisture decreases in summer by 15 percent to 20 percent of the present value. This increases the frequency and severity of drought.

Fortunately, this is an area of the world with excellent historical climate data well suited for evaluation trends during a time of known rise in greenhouse gas concentrations.

Figure 7.8 shows the past century's mean annual temperatures for the area. The plot shows relatively large variability from one year to the next. Temperatures rise from the late 19th century to the mid-1930s but have generally cooled since then. The linear temperature change either from the beginning of the record to the present or in the last half of this century is not statistically significant. Despite

Figure 7.9
20TH-CENTURY MIDWESTERN U.S. PRECIPITATION

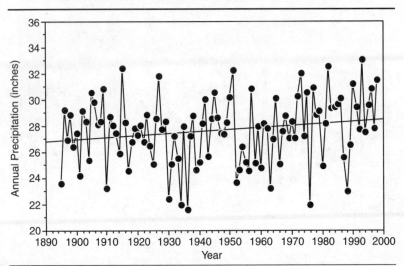

NOTE: The history of Midwestern precipitation is opposite what the IPCC predicted.

the expectation of warming given the buildup of greenhouse gases over the past centuries, the mean annual temperatures in the central United States have remained unchanged.

When the data are broken up into seasons, the summer temperatures decline slightly during the recent five decades but increase slightly overall. Winter temperatures show a small, but not significant, increase over the entire 1895–1996 period, but no change during the most recent 50 years. Ultimately, the IPCC's forecast for central North America does not find much support in the observational temperature records of the past century or half-century.

Analysis of precipitation records also shows little support for the prediction of increasing precipitation in winter and decreasing precipitation in summer. The total annual precipitation in the region shows a rise of about 50 millimeters (1.97 inches) over the past century, and the trend is statistically significant (Figure 7.9). Over the entire century of records the summer and winter upward trends were similar, but during the most recent 50 years, the rate of the summer increase has been more than 50 percent greater than the rate of winter. This is completely opposite the IPCC's prediction.

Figure 7.10
MIDWESTERN U.S. SNOW DAYS

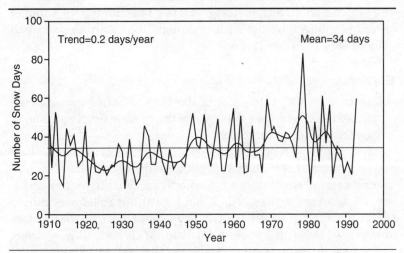

SOURCE: Hughes and Robinson, 1996.
NOTE: Duration of snow cover is increasing in the upper Great Plains, contrary to GCM predictions.

The observational data show that the Great Plains have become slightly wetter over the past half-century, despite model predictions for warming and drying. Given these empirical findings, it is not surprising that Hughes and Robinson, in a 1996 paper in the *International Journal of Climatology*, found a statistically significant *increase* in snow cover duration in the central United States (Figure 7.10). They found that the upper Great Plains region has gone from 29 days with snow cover greater than 7.6 cm (3.0 inches) early in the record to 38 days in recent decades. Although the increase is most pronounced in the autumn season, with little change in spring, the snow data certainly do not show the reduction the model simulations would lead a person to expect.

The analyses of historical drought patterns in central North America bring us once again to the dilemma that is always at the heart of the greenhouse debate. The numerical model simulations would have us expect increasing temperatures, decreasing summer rainfall, and a decrease in the soil moisture levels. But we can check these expectations against reality because the region has excellent temperature, precipitation, and drought records. In fact, none of the expected seasonal patterns are found in the observations. Temperatures in the area are not rising, rainfall rates are up in summer instead of winter, and the leading drought indicators show a trend to increasing moisture levels.

The Future

This discussion of drought in the United States captures the essence of the greenhouse debate. Are the models simply wrong, or is something else at work here?

Following the severe drought in the United States in 1988—which initiated much of the current greenhouse ballyhoo—many climate scientists set out to determine the underlying cause of the dry conditions. Kevin Trenberth of NCAR and his various colleagues carried out the most comprehensive work. They found that the 1988 summer drought was related to low rainfall in April and May, a strong anticyclone (high-pressure system) over the area with a northward displacement of the jet stream, cool water in the equatorial eastern Pacific (called La Niña), warm water in the Pacific from 10° north to 20° north, and a northward shift in the intertropical convergence zone.

Those authors argued, largely from theoretical considerations, that sea surface conditions in the Pacific had altered the atmospheric circulation system in a way that discouraged precipitation in the central United States. They further posited that the antecedent dryness created a positive feedback loop in which drought leads to increased surface heating, which then intensifies drought itself. A similar process largely contributed to the severity of the dust bowl droughts in the Great Plains nearly 70 years ago.

The argument may be true in a theoretical sense, but other factors can overwhelm it. In 1999, the La Niña–Midwestern drought link was broken as generally moist conditions prevailed in the midcontinent. The next chapter considers the statistical relationships between El Niño and U.S. moisture anomalies in more detail and finds little evidence for a very strong signal over most of the nation, despite reams of popular articles saying the opposite.

With respect to the greenhouse issue, some GCMs, such as the one published by Meehl and others in *Nature* in 1996, calculate that an increase in the concentration of carbon dioxide leads to more

El Niño (warm Pacific) events; indeed, Trenberth made the same argument in his 1998 congressional testimony. But if his theoretical arguments are correct, then there should be more *infrequent* drought in the central United States. If the statistical (vs. the theoretical) analyses are right, it will not matter much one way or another.

Finally, the prediction of increasing droughts in the central United States may be "corrected" by adding other variables to the computer models—that may not be warranted—such as sulfate aerosols. The IPCC claimed in its 1996 report that this lessened the reduction in soil moisture resulting from reduced warming, lower potential evapotranspiration levels, and enhanced summer precipitation levels.

The drought story ends on a very positive note, at least for those of us not involved in scaring people with greenhouse predictions. The long-term trends in soil moisture conditions appear to be toward increased moisture and less drought. The models of gloom and doom, and their proponents, were *wrong*.

Severe Floods

The concept of more intense and extreme rainfall's producing greater and more frequent flood events entered the popular mind with the publication of a 1995 *Nature* paper by Tom Karl. The increase Karl and colleagues wrote about was specifically in rains of two to three inches in 24 hours—heavy rains, to be sure, but hardly disastrous. Vice President Al Gore popularized (nay, *stretched*) their fearfulness by referring to such events as "torrential rains."

The linking of global warming and heavy rainfall derives from the fact that as the atmosphere warms, evaporation of surface water increases the amount of water vapor in the air. Warmer air, coupled with a moister atmosphere, produces conditions conducive for the development of heavy precipitation.

Missing from disaster scenarios, however, is another fact: In most parts of the United States (and the world), more precipitation is beneficial. In most of the United States, during the summer, total evaporation greatly exceeds total precipitation. Hence a moisture deficit exists. The lack of moisture stresses plants, including crops. Most farmers, therefore, welcome more summertime rainfall. And so does almost everyone else, because increasing population density puts more pressure on water supplies as the need for water in the

Figure 7.11
RAINFALL INCREASE IN THE UNITED STATES, 1910–90, IN INCHES

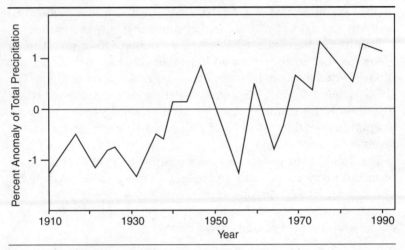

SOURCE: Karl, Knight, and Plummer, 1995.

NOTE: This figure shows change in the percentage of U.S. rain coming from storms of two inches or more in 24 hours. In reality, there was no change in storms of more than three inches in this study. The total change in rainfall works out to less than one inch.

industrial, commercial, and private sectors rises. Any increase in the amount of precipitation helps to meet the growing demand.

But Gore and others make this positive effect sound catastrophic. In fact, Karl's paper merely showed that rainfall of more than two inches per 24 hours had increased in the United States by an amount so small that no one (aside from climatologists) could have noticed.

At the time the paper was published, Karl told Michaels that he found no statistically significant increase in rainstorms of three inches or more. Storms of less than three inches are hardly "torrential." More remarkably, the percentage of total rain from the two-to-three-inch storms that normally falls on the United States is a mere 11 percent of the total annual rainfall of 34 inches, or 3.74 inches. When Karl's data begin, around 1910, that type of rain made up 9 percent of the then-average 31 inches, or 2.79 inches. In other words, the *difference* that has evolved in the century is a mere 0.95 inches of more rain coming from these storms in an average year.

Yet based on Karl's results, a January 1997 U.S. Department of Commerce press release claimed "flooding rain" had increased 20 percent. Karl had found a change of 2 percent, *not* 20 percent. Why did Commerce claim 20 percent? No, it was not a typo. By their logic, since at the beginning of the record 9 percent of all rain came from such storms and by the current decade it is 11 percent, they divided 2 percent by 10 percent (the average of 9 and 11) to get 20 percent!

In work related to Karl's, Lins and Slack of the U.S. Geological Survey looked directly at streamflow from hundreds of U.S. streams and rivers, publishing their results in *Geophysical Research Letters* in 1999. They searched for statistically significant trends in both low (drought) and high (flood) flow categories, and they also evaluated flow in different percentiles (90th percentile, 70th, etc. . . . down to 0) to get an overall picture of the distribution of changes. They found a strong tendency for a decrease in the frequency of low flow (i.e., less drought) and no change in high flow (flood). They wrote:

> The pattern indicates that baseflows are increasing (which suggests that drought is decreasing), median or average stream flow is increasing, but annual maximum flows (including floods) are neither increasing nor decreasing. Hydrologically, the nation appears to be getting wetter, but *less extreme* [emphasis added].

Lins and Slack used only those stream gauges upstream of which the water flow was undisturbed. This restriction ensured that alterations of the watershed or the stream channel did not influence the data they were using—a good thing, since those types of alterations are undoubtedly giving rise to the public's perception of increasing floods. Think about it: As land use increasingly becomes more urban, the surface becomes more and more impervious and runoff increases. And as streams and rivers are channeled and diked, the natural system of checking floods—flood plains—is eliminated and all the water is forced to flow in narrow stream channels. This has the effect of increasing downstream water levels, often to heights we have never seen before. The result? Seemingly record floods, despite the current waterway's bearing no resemblance to the one of the past against which the records are being compared.

On a larger scale, the IPCC reports a slight (1 percent) upward trend in precipitation over the globe's land areas during the 20th

135

century, though they note that since about 1980, precipitation has been relatively low. On regional scales, of course, this trend has exhibited variability. As for a tendency toward increased flooding, the IPCC reports that no clear, large-scale pattern has emerged.

Temperature Extremes

During February 3–5, 1996, a large number of long-running U.S. climate stations set their all-time low temperature records for any date. The National Climatic Data Center's station in Tower, Minnesota, for example, registered $-51°C$ ($-60°F$). The unurbanized record at Charlottesville, Virginia, which begins in 1895, hit its all-time low at $-24°C$ ($-11°F$), a phenomenally cold temperature considering that it is not a high-elevation location at latitude 38° North, and that it is near a warm ocean. Across much of the nation, it was bitter cold. That week, speaking at a New Hampshire elementary school, President Clinton blamed the cold weather on global warming.

That is right—he claimed that warming causes cooling. Extreme temperatures are often cited as yet another likely consequence of global warming. Extremes have direct consequences on human health and mortality, as well as on energy usage and agriculture. To assess such effects, a careful understanding of the way future climate change may impact the magnitude and frequency of extreme temperature outbreaks is necessary.

High Temperatures

Eleven years ago, James Hansen predicted that by the 2050s, Washington, D.C., would see an increase in days with temperatures of more than 95 degrees from a normal of six to nearly 50. Similar statistics were calculated for other cities across the country. Those forecasts served as the basis for a memorable Sierra Club television advertising campaign in 1990 featuring Meryl Streep, William Shatner, and some actors from the popular 1980s yuppie melodrama *Thirtysomething*.

A closer look at the numbers, however, proves the statistics to be based upon "facts" that are known to be wrong. Hansen arrived at the value for the number of extremely hot days by simply adding the amount of model-projected warming to the normal distribution of maximum temperature values occurring during the summer (Figure 7.12). But that practice does not consider the nature of the observed temperature changes or those the climate models project.

136

Figure 7.12
A MISGUIDED FORECAST OF FUTURE TEMPERATURE DISTRIBUTION

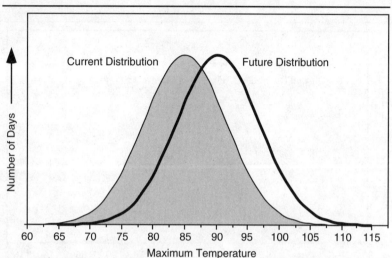

NOTE: If the forecasts of future temperature increases are simply added to the current temperature distribution, the number of extremely high temperatures greatly increases. But the result is misleading.

Karl's 1997 *Scientific American* paper states that GCMs predict the variance of temperatures will decline in the future—that is, temperatures will be confined to a tighter range. Indeed, this effect is evident already on an annual basis. Winter temperatures are rising at more than twice the rate of summer temperatures, which must narrow the annual range.

Therefore, adding the forecast warming to the currently existing temperatures is incorrect. Greenhouse warming is increasing night temperatures more than day readings (which therefore means more warming in the winter, when nights are longer, than in the summer). For this reason, it is necessary to adjust the temperature forecasts for these changes in the day–night cycle. Using this equation, the number of extreme events shows little change and, in some cases, may in fact actually decrease in number (Figure 7.13); Table 7.1 provides some evidence that this is happening.

To assess any trends that might have contributed to the deaths in Chicago during the brief but intense heat wave of July 1995,

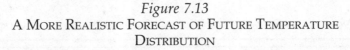

Figure 7.13
A MORE REALISTIC FORECAST OF FUTURE TEMPERATURE
DISTRIBUTION

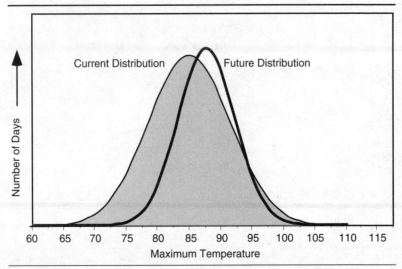

NOTE: If the forecasts (and observations) of decreases in temperature variance are included, a future warming may significantly affect the number of extremely high temperatures. Whether or not this occurs depends on whether the variability of the data shrinks more than the predicted mean warming.

Karl went back and examined the long-term trends of maximum "apparent temperatures" (temperatures plus the effect of humidity). Writing in the 1997 *Bulletin of the American Meteorological Society*, Karl reported that instrument changes and urbanization had the greatest influence in the apparent temperature; when he corrected for these effects, he found no discernible trend.

David Easterling, writing in *Science* in 1997, found that most of the increase in global temperatures has been occurring during the winter and at night. Summer maximum temperatures in the Northern Hemisphere showed no significant trend. This is another fact that is apparently overlooked when one merely adds the forecast temperature change to the currently observed maximum temperatures.

The authors' own 1998 *Climate Research* study looked at the occurrence of record-setting daily high temperatures across the United

States, China, and the former Soviet Union, finding that there was an overall tendency for daily record high temperatures to be set earlier in the record. The IPCC drew the same conclusion in its 1996 report, which states that "widespread significant changes in extremely high temperature events have not been observed."

Low Temperatures

According to the *Vital Statistics of the United States*, more than twice as many people die from exposure to extreme cold weather as do from exposure to extreme heat. During an arctic outbreak that reached into the deep South during Christmas 1983, some 150 people died. The low temperature in Orlando, Florida, was 20°F with a wind chill factor of −6°F. Many of the victims died of hypothermia in their unheated homes.

This killer air mass had formed over Siberia, a dark, dry frozen expanse of land where little acts to brake the escape of radiation into space, and temperatures routinely drop to −40°C (−40°F). Siberian air masses travel over the pole and when conditions are right, they can drop deep into the southern United States, often with dire consequences. Fortunately, as the carbon dioxide concentration in the atmosphere increases, these are the air masses that should warm the most.

Recall that greenhouse warming should be amplified in the coldest, driest air masses. Indeed we now know that the largest temperature increases have been occurring during the winter over Siberia and northwestern North America. The warming of these cold polar air masses reduces the frequency of extremely low temperatures. This lessens the severity of cold air outbreaks and reduces the number of deaths resulting from these air masses' excursions into the lower latitudes.

Storms

Hurricanes

There is nothing like a good hurricane to generate weather and climate angst. News organizations love a hurricane because it usually takes days of not completely predictable meandering before it splatters upon the densely populated Atlantic or Gulf Coasts. From its inception to the obligatory Sunday news coverage of the

139

survivors in church, a good hurricane provides a solid week of news. (Considering recent appellations such as Georges and Babs, we are thankful that "Kennedy" is not in the catalog of hurricane names.)

Global warming fear-mongers love hurricanes because they provide free advertising. If they can somehow ensconce in the public consciousness the belief that global warming will make hurricanes worse, more frequent, or both, they have won a powerfully emotional point.

The media are not immune. "The Hot Zone: Blizzards, Floods & Hurricanes: Blame Global Warming," proclaimed *Newsweek*'s January 22, 1996, cover, which showed a man walking through the white-out of a snowstorm. Clearly, global warming–related extreme weather had entered the popular consciousness. The IPCC remained conservative in its discussion of the issue, stating in the Technical Summary of its 1996 report, "Although some models now represent tropical storms with some realism for present day climate, the state of the science does not allow assessment of future changes." In the body of the text, the IPCC wrote, "In conclusion, it is not possible to say whether the frequency, area of occurrence, time of occurrence, mean intensity or maximum intensity of tropical cyclones will change."

Will an increased concentration of atmospheric carbon dioxide result in more frequent and more intense hurricanes? In 1986, Kerry Emanuel published a paper in the *Journal of the Atmospheric Sciences* dealing with air–sea interactions and hurricane activity. He showed that if the sea surface temperature falls below about 26°C (79°F), intense hurricanes would become a physical impossibility. Further, the intensity of a hurricane has a well-defined upper limit that is governed, in part, by sea surface temperatures. In simple terms, a warmer sea surface could theoretically increase the upper limit of a storm's intensity. Only a few storms actually approach this theoretical upper limit, but those storms turn out to be the most destructive and dangerous.

The following year in *Nature*, Emanuel published another, much more publicized article on the dependence of hurricane activity on climate. He showed that a computer model perturbed by an increase in greenhouse gases could increase the sea surface temperature, thereby increasing the theoretical upper limit of storm intensity. This is not to say that the world would see more hurricanes or even more intense hurricanes, but only that the theoretical upper limit of intensity could increase. For a 3°C (5.4°F) increase in sea surface temperatures, the potential destructive power of storms approaching this theoretical limit, as measured by the square of the wind speed, could increase by 40 percent to 50 percent.

A third Emanuel paper, again in *Journal of the Atmospheric Sciences,* in 1988, further developed the linkage between the upper limit of storm intensity and warming in the lower atmosphere. He wrote that with a warming of the sea surface of 6°C (11°F) to 10°C (18°F) (and with conditions in the lower stratosphere held constant), a supersize, ultrapowerful "hypercane" becomes a theoretical possibility. Emanuel presented these ideas to a more popular audience in a paper he prepared for *American Scientist.* There, the general public learned about the hypercane and its link to warmer conditions.

Other theoretical results are in agreement with the notion that hurricane activity will increase due to the buildup of carbon dioxide. Writing in *Geophysical Research Letters,* Ryan et al. (1992) concluded that the area affected by hurricanes would increase. A computer modeling study by R. J. Haarsma et al. found that both the frequency and intensity of tropical cyclones would increase in a warmer climate; their paper was published in 1993 in *Climate Dynamics.*

Increased sea surface temperatures in the future would decrease the central pressure of the storms (and therefore increase their intensity), and tropical cyclones would generate more rainfall, wrote Evans et al. in a 1994 *Journal of Climate.* In *Science,* Tom Knutson's computer models for a CO_2-warmed world showed hurricanes with lower pressures and increased winds. J.-F. Royer et al. (1998) examined the effect of increased greenhouse gas concentrations on climate and found an increase in the number of tropical cyclones in the Northern Hemisphere and a decrease in the Southern Hemisphere in a paper published in *Climate Change* in 1998.

Other scientists, though, are producing calculations that lead to a different conclusion—that elevated carbon dioxide levels might even *decrease* tropical cyclone activity. In the journal *Tellus* in 1995, Bengtsson used a global climate model and found a reduction in hurricanes, particularly in the Southern Hemisphere, when the climate warmed. One year later, again in *Tellus,* Bengtsson and others published another paper entitled, "Will Greenhouse Gas-Induced

Warming over the Next 50 Years Lead to a Higher Frequency and Greater Intensity of Hurricanes?" He apparently found the result so striking that he italicized it in the paper's uncharacteristically passionate abstract: ". . . the number of storms is significantly *reduced*, particularly in the Southern Hemisphere . . . most tropical storm regions indicated reduced surface wind speeds and a slightly weaker hydrological cycle." Their high-resolution numerical simulations coupled with an ocean-atmosphere model showed that greenhouse-induced changes would weaken the subtropical high-pressure systems that dominate the tropics. Reducing these changes drops the strength of the trade winds, which weakens hurricanes. It would also strengthen the upper-level west winds in the vicinity of hurricane development, which further weakens hurricanes.

When compared with the global distribution and seasonality of hurricanes today, they found no changes for a doubling of greenhouse gases. But the number of hurricanes in the Northern Hemisphere fell from 56.2 storms per year in the present-day climate simulation (the observed value from 1958 to 1977 is 54.6) to 42.0 storms per year in the doubled CO_2 case. In the Southern Hemisphere, the number of hurricanes dropped from 26.8 in the present-day model run (24.5 is the observed value) to only 11.6 storms per year. Their results on intensity were less conclusive, but they did find a tendency for reduced wind speeds in the doubled CO_2 model simulations.

Again, the level of disagreement in the theoretical scientific literature prompts an adjudication by the observed data. While the computer modelers have been busy conducting theoretical experiments on the linkage between hurricanes and warmer conditions, other atmospheric scientists have been examining the historical record to determine the relationship between observed temperatures and observed hurricane activity as well as trends in hurricane activity.

In 1990, noted hurricane scientist William Gray (who makes the most reliable season-in-advance hurricane predictions) published an important article in *Science* dealing with landfall of intense hurricanes in the United States and its relation to rainfall in West Africa, showing that hurricane activity over the period 1970–87 was less than half of that observed for the period 1947–69. Even though the largest greenhouse gas increases are in the latter period, hurricane activity was decreasing overall. Drought in West Africa is negatively correlated with Atlantic hurricanes, probably because the "seedling" disturbances that ultimately become Atlantic storms originate as clusters of thunderstorms over West Africa, some of which originate in Indian monsoon, which El Niño weakens, according to a 1999 *Journal of Climate* article by Chris Torrance and Peter Webster.

In *Meteorology and Atmospheric Physics*, Idso and Balling (1990) analyzed hurricane data for the central Atlantic, U.S. East Coast, the Gulf of Mexico, and the Caribbean Sea for the period 1947–87. They found that "there is basically no trend of any sort in the number of hurricanes experienced in any of the four regions with respect to variations in temperature." The number of hurricane-days (one hurricane on one day equals one "hurricane-day"; two on one day equals two "hurricane-days," etc.) was negatively related to the Northern Hemispheric temperatures; warmer years produced the lowest number of hurricane-days, while cooler years had more than the average number of hurricane-days.

Nor did the number of storms within the various intensity classes correlate with the hemispheric temperature values (Figure 7.14). They examined the trends with different intensity classes and concluded, "For global warming on the order of one half to one degree Centigrade, then, our analyses suggest that there would be no change in the frequency of occurrence of Atlantic/Caribbean hurricanes."

Later, the appropriately named hurricanologist Christopher Landsea and coworkers reported in *Geophysical Research Letters* (1993) that the intensity of Atlantic hurricanes has been decreasing since the middle of this century. Landsea carefully screened his data to remove known biases, and the decreasing trend remained an identifiable pattern in the intensity estimates. Another article by the same group reconfirmed that hurricane frequency and intensity were not increasing over the past five decades. They examined Atlantic hurricanes from 1944, when aircraft reconnaissance began in the Atlantic, to the near-present. They found that "a long-term [five decade] downward trend continues to be evident primarily in the frequency of intense hurricanes. In addition, the mean maximum intensity [i.e., maximum wind, averaged over all cyclones in a season] has decreased." A plot of the mean intensity (Figure 7.15) clearly shows this downward trend during a time of greatest buildup of greenhouse gases.

The downward trend in intensity of Atlantic hurricanes over the past five decades raises a question about trends prior to the 1940s,

143

Figure 7.14
ATLANTIC/CARIBBEAN HURRICANES AND NORTHERN
HEMISPHERIC TEMPERATURE VARIATIONS

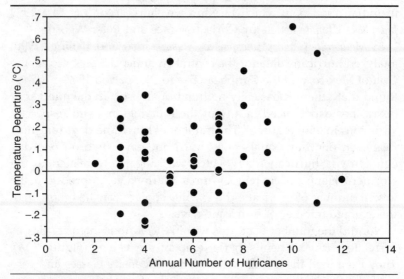

NOTE: This figure shows the number of Atlantic/Caribbean hurricanes vs. Northern Hemispheric near-surface air temperature departures from normal.

when data are less reliable. Karl et al. examined records of the number and intensity of hurricanes that reached the continental United States over the past century and found that the number of hurricanes making landfall in the United States decreased from the 1940s through the 1980s (Figure 7.16), but that over the entire 20th century, there was no overall trend. This was one of the first papers published in a new, federally funded climate change journal with the ominous title of *Consequences*.

In a 1996 *Journal of Climate* paper that was very important but received no public attention, Elsner et al. looked at Atlantic hurricanes from 1896 to 1990 and noted a drop in those originating in the low-latitude tropics beginning in the 1960s. Hurricanes originating there tend to be more severe than those that originate elsewhere in the Atlantic Basin. Chan and Shi (1996), in *Geophysical Research Letters*, examined the record of typhoons and tropical cyclones from 1959 to 1994 in the western North Pacific and found a decrease in

144

Figure 7.15
WIND SPEEDS OVER TIME IN ATLANTIC/CARIBBEAN HURRICANES

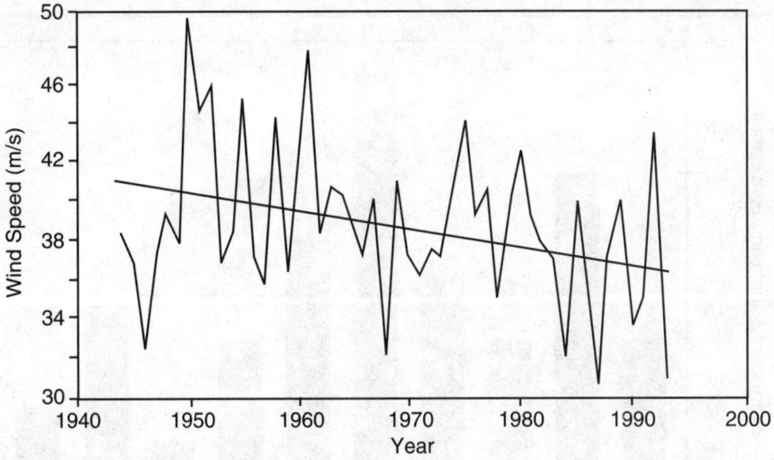

NOTE: This figure shows time series of Atlantic basin mean intensity (meters per second) as determined from maximum sustained wind speeds of all hurricanes.

storms from 1959 to the late 1980s and a trend upward since (through 1994). But when the most recent years—after 1994—are included, there is no significant change at all. Nicholls et al., in *Meteorology and Atmospheric Physics* in 1998, analyzed satellite records of tropical cyclone activity in the Australian region and found a decline over the past three decades.

Finally, in a 1998 *Bulletin of the American Meteorological Society,* Bove et al. examined Gulf of Mexico hurricane landfalls from Florida to Texas for the period 1886 to 1995 and concluded that "there is no sign of an increase of hurricane frequency or intensity in the Gulf of Mexico at this time." Their final sentence states, "Fears of increased hurricane activity in the Gulf of Mexico are premature." The use of the word "premature" says something about the authors' preconceptions; in fact, the fears are "misplaced" or "wrong," not merely premature.

In addition to studies examining precise aspects of a certain scientific question, there are "overview" studies that review the available science to draw broad conclusions. In the first scientific assessment

Figure 7.16
NUMBER OF HURRICANES MAKING LANDFALL IN CONTERMINOUS
STATES, 20TH CENTURY

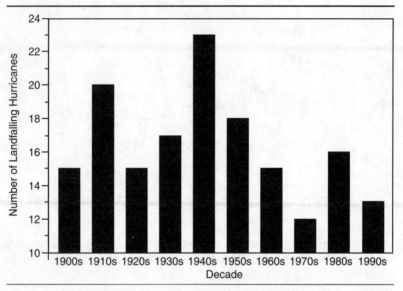

from the IPCC in 1990, the "Policymakers Summary" states, "Climate models give no consistent indication whether tropical storms will increase or decrease in frequency or intensity as climate changes; neither is there any evidence that this has occurred over the past few decades." Four years later, in the *Bulletin of the American Meteorological Society*, Lighthill et al. published a major paper using two different basic approaches, both of which led to the conclusion that "even though the possibility of some minor effects of global warming on tropical cyclone frequency and intensity cannot be excluded, they must effectively be 'swamped' by large natural variability."

Lighthill et al. examined a widely accepted list of conditions permitting hurricane formation and development, such as a sea-surface temperature above 26°C (79°F), distance from the equator of at least 5° of latitude, and fairly high relative humidity levels surrounding the storm. Other entries on the list dealt with more complex conditions such as the vertical temperature structure of the atmosphere and horizontal rotation of the system. Their analyses led them to conclude that we should not expect any direct effects of changing sea surface conditions on hurricane frequency and intensity. Lighthill et al. also examined the empirical records of hurricane activity since 1944 in the Atlantic and since 1970 in the Pacific. The authors noted great year-to-year variability in the hurricane data, but they could not find evidence linking hemispheric temperatures to variations in hurricane activity.

The 1996 IPCC report stated, "Although some models now represent tropical storms with some realism for present day climate, the state of the science does not allow assessment of future changes." It went on to summarize, "[I]t is not possible to say whether the frequency, area of occurrence, time of occurrence, mean intensity or maximum intensity of tropical cyclones will change." Similarly, Karl's *Scientific American* article concluded that "overall, it seems unlikely that tropical cyclones will increase significantly on a global scale. In some regions, activity may escalate; in others, it will lessen." The most recent (1998) review article was published by Henderson-Sellers et al. in the *Bulletin of the American Meteorological Society*. It concluded, "There are no discernible global trends in tropical cyclone number, intensity, or location from historical data analyses." The authors also stated that the bulk of the evidence suggests little change will occur in hurricane activity over the next century, considering the expected changes in atmospheric composition.

Given this voluminous scientific literature, the glib, even sensationalistic nature of the *Newsweek* cover conflating global warming and hurricanes is disappointing, to say the least.

Extra-Tropical Storms

Extra-tropical storms, or cyclones (areas of low-surface air pressure that form outside the tropics) draw their energy from a different source than do tropical systems (hurricanes, etc.). A key component of the climate systems, they are critical to any discussion of climate change. Although the general term "cyclone" applies to almost any weather feature with "cyclonic" (in the Northern Hemisphere, counter-clockwise) winds, this discussion focuses on large-scale middle and high-latitude cyclonic storms such as those represented on surface weather maps.

Cyclonic storms are responsible for most of the significant weather events outside of the tropics. Low-pressure systems and their related fronts produce most of the precipitation, especially in the autumn, winter, and spring. With the exception of hurricane-produced rains, severe summer thunderstorm complexes, including tornadoes, are almost always linked to some type of cyclonic disturbance, even though it may be much weaker than one in the winter. All major (and most minor) snowstorms are cyclone-related. Thus, cyclones include almost all major nontropical precipitation producers.

Some people have suggested that global warming will increase the frequency and violence of these types of storms. The popular vision of the warmed world of the future, as the infamous *Newsweek* cover epitomizes, is a climate of stronger blizzards, more forceful winds, and more intense tornadoes. But this picture is based on an incomplete understanding of the atmospheric dynamics that are involved in the creation and evolution of midlatitude cyclones. Mid-latitude storms derive their energy and sustenance from changes in the atmospheric circulation (large-scale wind systems) at all levels of the atmosphere up to and including the jet stream (about 30,000 feet).

Surface cyclones are regions with converging surface winds—winds that blow toward a common center. For the cyclone to main-tain its low pressure, as the air spirals into a storm at the surface and rises (producing clouds and precipitation), a mechanism aloft must remove the air being added from below. This phenomenon occurs in the upper atmosphere near the level of the jet stream. A strong jet is needed above a strong cyclone to generate divergence (air flowing away to different locations) aloft that at least compen-sates for the convergence at the surface. Any discussion of future changes in cyclonic systems and the resultant consequences requires an analysis of the future upper atmosphere circulation changes.

Circulation Changes

The global wind patterns in the upper atmosphere change on a daily, seasonal, and annual basis. Westerly winds surround the poles of each hemisphere, producing, as viewed from a location over each pole, a "circumpolar vortex" that effectively isolates the cold polar air from the warmer tropical air. Moving south to north over the hemisphere, the area of the strongest temperature contrast is where the jet stream, or the region of maximum winds, is located.

While tropical temperatures remain roughly constant throughout the year, high latitude temperatures change significantly from summer to winter. As the polar latitudes warm, the north–south temperature contrast weakens, the circumpolar vortex (and the pool of cold air) contracts poleward, and the jet stream slackens. These seasonal changes correspond closely with changes in the frequency, intensity, and tracks of surface cyclones. In winter and spring, when intrahemispheric temperature contrasts are strongest, cyclones track farther south throughout the hemisphere and are more intense. Summer sees fewer, weaker cyclonic storms with tracks displaced to the north.

How will increasing trace gas levels change the atmospheric circulation? Most GCMs predict that the greatest warming will occur over the high latitudes in winter with comparatively little warming in the tropics. So, according to these simulations, the future contrast between the polar latitudes and the tropics should lessen, particularly in winter, producing a weaker jet stream, a more contracted vortex, with fewer, and/or less powerful cyclones. In short, the future atmospheric circulation should be less "winterlike," with fewer intense storms.

Interestingly, atmospheric GCMs, including the generation of models that served as the basis for the 1992 Rio climate treaty and the 1996 IPCC report, do not explicitly incorporate cyclones. Typical midlatitude storm systems span hundreds of miles but are nevertheless smaller than the resolution scale of most global GCMs. It is possible, however, to run higher-resolution models over limited spatial areas and to produce more regional experiments that explicitly include cyclones. For example, Beersma published a 1997 paper in *Tellus* in which he examined changes in North Atlantic cyclones between atmospheres with current and doubled carbon dioxide levels. He noted "a shift in the distribution of depression [cyclone] strength towards weaker depressions. As a result, the number of storm events. . .decreases in most areas." He further stated that it was "impossible to conclude that the modeled differences were the result of the CO_2 doubling because they were small compared with the background natural variability." In another study of cyclones, published in 1995 in the *Journal of Climate*, Zhang and Wang ran the NCAR model for trace gas conditions of 1990 vs. projections for 2050 for the Northern Hemisphere. They detected (1) a decline in the number of cyclones, (2) a weakening of cyclones in their formation regions, and (3) fewer coastal cyclones forming, particularly in winter. These U.S. Atlantic coastal storms, commonly called "nor'easter," are responsible for considerable damage and very heavy snow in the urban corridor of eastern North America.

149

Using the United Kingdom Meteorological Office (UKMO) GCM, Murphy and Mitchell, in *Journal of Climate* in 1995 found that "increased westerlies which coincide with an increase in the strength of the Atlantic storm track are associated with an increase in the south-to-north temperature gradient between 30° N and 60° N."

Their model forecasts more North Atlantic cyclone activity along with stronger circulation arising from a stronger north–south temperature gradient, but in doing so it contradicts both theory and most other GCMs, which call for a weaker temperature gradient and a decrease in the strength of the cyclonic storm track. Nor do observations support the UKMO results: In a 1997 article in *Journal of Climate*, Serreze and others identified a significant *decrease* in the number of cyclones in the North Atlantic, using observed data rather than model results.

With the exception of the UKMO model, which runs counter to other forecasts, both projections and reality indicate that the number and intensity of storms decrease as the greenhouse effect is enhanced. There is little basis, in theory or in fact, for seeing a rise in the frequency or intensity of midlatitude storms as symptomatic of a global warming response. Global warming may very well produce the exact opposite.

Circumpolar Vortex Changes

As cyclone frequency and strength are determined by the strength of the westerly vortex surrounding the pole, perhaps it might be useful to look explicitly at this "cause" of storms. Robert Davis and Stephanie Benkovic, in a 1992 article in *Theoretical and Applied Climatology*, plotted changes in the January circumpolar vortex in the mid-troposphere (approximately 18,500 feet) and found that, from 1947 to 1990, the January vortex expanded significantly. That is the opposite of what we would expect in a global warming scenario. In a separate 1993 analysis in *Journal of Climate*, Aaron Burnett found the same thing, corroborating Davis's original analysis.

Later work by Davis and others, presented to the American Meteorological Society in 1997, uncovered *no* statistically significant contraction of the vortex in any month, but significant expansion in four of the 12 months from 1947 to 1995. Conversely, Angell's 1999 study in the *Journal of Geophysical Research* notes a significant 1.4 percent per decade contraction of the circumpolar vortex at a higher level (around 25,000 feet) over the period 1963–97. This difference between atmospheric behavior at different upper altitude levels may be real, or it could be a function of different time periods used in the analyses.

Cyclone Frequency

Cyclone histories typically rely on cyclone counts and the tracking of low-pressure systems through some type of geographic grid. Early research by a host of independent investigators documented significant changes in the year-to-year variability in cyclone counts, intensity, and tracks. In a more recent (1996) study in the *Journal of Geophysical Research*, Lambert analyzed the yearly counts of intense cyclones in the Northern Hemisphere and detected no trend from 1899 to 1970, but noted a statistically significant increase from 1970 to 1991 over the Atlantic and Pacific oceans. Lambert then argued that the stronger Pacific storms are related to lower water temperatures in the basin. In a more regional-scale study in the 1998 *Journal of Climate*, Angell and Isard detected a significant increase in the number of strong storms in the Great Lakes region. They also found that more storms occurred in cold years than in warm ones.

So observations show that more frequent cyclones are associated with lower temperatures. This supports the theory that global warming should produce fewer storms and fewer intense storms, owing to a decline in the hemispheric temperature difference between polar and tropical latitudes that provides the jet stream energy needed to spool up cyclones.

High winds are usually associated with cyclones. In Switzerland, Schiesser et al. examined 100 years of wind data in which they detected trends toward less windy conditions and fewer storms. Published in 1997 in *Theoretical and Applied Climatology*, their report suggested that this response could indeed be evidence of a global warming signal.

A "blizzard" is a peculiarly fatal combination of wind, cold, and snowfall created by cyclones. Though the definition varies form place to place (it does not have to be as cold in Washington, D.C., as it does in Bismarck, North Dakota, to qualify), the bottom line is the same: severe whiteout conditions that result in spatial disorientation and death to people who are otherwise well-dressed for their climate.

151

In the winter of 1996, after a spate of strong snowstorms and cold outbreaks, several government officials, with the support of a handful of climate scientists, contended that blizzards, too, could be related to global warming. NASA's James Hansen championed the idea in an interview in the infamous January 22 *Newsweek*. Hansen's argument was especially surprising since most GCMs indicate that the formation regions of coldest air are projected to warm more than any other place on the planet. Neither then nor since has any scientist logically defended the hypothesis that warming implies cooling.

The blizzard contention is similarly ill founded. Along the East Coast of the United States (and in Asia), strong snowstorms require three conditions to form: a deep wedge of cold air in place, a strong coastal cyclone, and sufficient moisture. Lack of any one ingredient or a slight change in the timing of the storm's arrival have changed many forecast major snowstorms into sleet or rain events.

For a blizzard to occur, a deep layer of cold air must typically be in place upon the arrival of the cyclone. Winds associated with strong coastal lows mix in warmer, oceanic air and moisture, which often raises air temperatures above freezing. But a sufficiently deep layer of very cold air is not easily eroded by the coastal storm winds, and the resulting precipitation falls as snow. Any warming of cold air would logically reduce the likelihood of blizzards in the future. Furthermore, the relationship between temperatures and total snowfall over most of the East Coast is negative, so warm winters have significantly less snow than cold winters, as recently published research by Davis et al. (1997) shows.

The second factor in snowstorm formation is the presence of adequate moisture. Since GCMs project a moister atmosphere, it is this primary feature that fuels speculation about more intense future blizzards. But sufficient moisture is almost never a limiting factor in snowstorm formation. Cyclones traveling along the coast have plenty of Atlantic moisture to tap into given the proper storm track. An already wet atmosphere has little room for additional moisture.

Before holding forth in *Newsweek* about a connection between East Coast snowstorms and global warming, writer Sharon Begley should have checked whether atmospheric moisture in the winter was increasing in that region. The standard measure for this is the history of "dewpoint" temperatures, which are the temperatures at which,

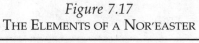

Figure 7.17
THE ELEMENTS OF A NOR'EASTER

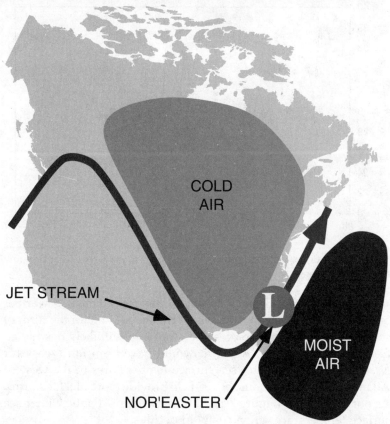

COLD
AIR

JET STREAM

L

MOIST
AIR

NOR'EASTER

NOTE: A snow "nor'easter" requires cold air from Canada, a dip in the jet stream, and moisture from the Atlantic Ocean. The latter two are often present in winter. Greenhouse warming should decrease the first factor.

when cooled, moisture will condense from the air (Figure 7.18.) The higher the dewpoint temperature, the wetter the air.

Michaels has published several papers, including two in *Geophysical Research Letters*, on historical trends in dewpoints. In winter over the eastern United States, there is simply no increase in dewpoint temperature, and therefore there is no increase in the amount of moisture available for snowstorms.

Figure 7.18
WINTER DEWPOINT TEMPERATURE TRENDS

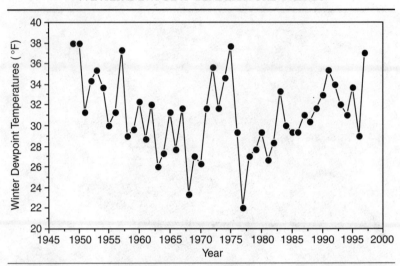

NOTE: Winter dewpoint temperature trends over the western Atlantic show no overall increase, contrary to speculations.

Hansen's contention was based on the concept that a warmer ocean evaporates more water, which then would fall in the form of snow. But remember, even a slight warming will likely destroy an East Coast snowstorm. Again, it would have been fairly easy for *Newsweek* to check winter sea surface temperatures in the western Atlantic Ocean, which must be rising to provide more fuel for storms. As our illustration (Figure 7.19) shows, western Atlantic winter sea surface temperatures are actually in decline.

The final key ingredient for snowstorms, strong coastal cyclones (or low-pressure areas), is slightly more complex. A 50-year record of coastal cyclones published by Davis and Dolan in 1993 in *American Scientist* reveals that, for the mid-Atlantic region, total storms are declining in frequency but strong storms are becoming more common. Could this be a global warming signal?

Major snowstorms require the presence of a strong "trough," or southward excursion, of the jet stream to mix polar cold and tropical warmth as far south as the Carolinas, which is a favored formation area for big nor'easters. However, Davis's analysis of the circumpolar vortex (which is the bottom of the jet stream) in warm vs. cold

154

Figure 7.19
DECLINING WINTER SEA SURFACE TEMPERATURES OFF THE MID-ATLANTIC COAST

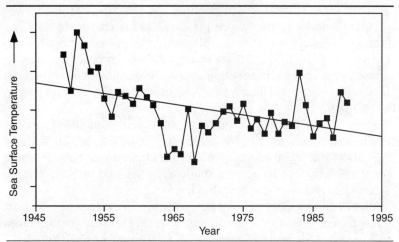

NOTE: Winter sea surface temperatures in the mid-Atlantic region are declining, which reduces the amount of moisture available for snowstorms.

winters shows no changes over the eastern United States. Furthermore, a decline in the coastal land–ocean temperature contrast implied by a warming of the coldest air masses, as we noted in the last chapter, reduces the likelihood of more future storminess.

In short, changing the greenhouse effect should produce a weakened jet stream that supports fewer cyclones and should warm the coldest air masses. In the midlatitudes, the warmer it is, in general, the less it snows. The greenhouse effect should be decreasing midlatitude snowstorms, something that simply has not been observed in any systematic fashion.

Tornadoes and Severe Thunderstorms

The great majority of the planet's tornadoes form in the United States, owing primarily to a coincidence of geography. The lack of analyses of long-term time series on tornado frequencies or strength is related to the substantial biases in the available data. Observed tornado frequencies have been increasing over time in the United States, but most of this trend is driven by improvements in our ability to detect tornadoes through technological advances and a

155

larger and more widely scattered populace with simple means to report tornado sightings to the authorities. There is no doubt that the new generation of weather radars, Doppler machines that can actually see the movement of raindrops, is detecting a large number of small tornadoes that more primitive radar could not see. The National Weather Service's Doppler network reached national coverage in the early 1990s. Expect some breathless report to appear in a few years about a dramatic increase in tornado frequencies; but that will be nothing more than an artifact of the new detection technology.

The upward frequency count coincides with equally significant declines in the number of tornado-related fatalities, which leads us to suspect that the frequency counts are increasingly biased. Nor is there much doubt that better tornado detection and public warming systems have saved innumerable lives.

Vice President Gore, who has been known to exaggerate or distort a few things, produced a wildly misleading statement on tornadoes in a June 8, 1998, press release. He said, "Tornadoes have killed 122 people this year, matching the annual record set in 1984." He made this assertion at a White House press conference condemning global warming, making him the first person we know of to unequivocally conflate tornadoes and global warming. His statement was illogical, disinformative, and wrong.

The total number of deaths recorded for the year ("annual record") in 1984 was indeed 122. But he used the word "set" as in "set a record." In fact, the death totals in 1984 were not anywhere near the all-time (or "record") high value. Figure 7.20 shows tornado deaths, a statistic that the U.S. Department of Commerce has kept since 1950. The most deadly year was 1953 because of a monster tornado in, of all places, Worcester, Massachusetts.

Perhaps Gore meant to say that 122 deaths is a record for only half a year; after all, he was speaking June 8. Wrong again. The vast majority of tornado deaths occur in spring and early summer; 85 percent of total annual tornado fatalities occur before July 1. There were *more* than 122 deaths in the first half of the year in 1953, 1957, 1965, 1971, and 1974. The only thing unusual about the tornadoes in 1998 was that that year marked the first time a president or vice president said anything so misleading about them.

Most tornadoes, and strong ones in particular, require a combination of warm, moist air near the surface, warm, dry air nearby, and

Figure 7.20
U.S. TORNADO DEATHS, 1950–99

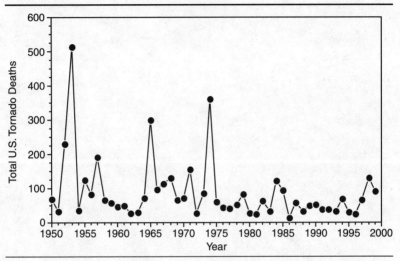

NOTE: The number of deaths from tornadoes in 1984 is not particularly unusual.

cold, dry air aloft. A strong cyclone and a strong jet stream are usually sufficient to meet these conditions. A weaker jet stream, as should occur with increasing levels of carbon dioxide, would therefore decrease the likelihood of tornadic activity, although this effect may be somewhat balanced by increases in the atmospheric moisture levels. All in all, an enhanced greenhouse effect should have little effect on the intensity or frequency of tornadoes.

8. An Ocean of Data: Sea-Level Rise and El Niño

So far, it has been pretty easy to demonstrate that greenhouse warming is likely to be at or below the lower limits of the range that dominated the discussion 10 years ago. The profound linearity of most climate change scenarios means that warming trends in the surface temperature established over the last three-plus decades may be quite indicative of the future (unless most every climate model has the mathematical form of greenhouse warming wrong). Further, the winter-to-summer half-year warming ratio is likely to maintain itself. The bottom line is a warming of around 1.5° to 1.7°C (2.7° to 3.0°F) in the winter half-year and 1.2° to 1.3°C (2.2 ° to 2.3°F) in the summer over the course of the next century, depending upon assumptions about the sun's behavior.

What is difficult is generating much public angst over these numbers. Once proponents of the catastrophic warming scenario are forced to concede that overall warming will not approach the values the models predicted 10 years ago, they usually substitute forecasts of disastrous flooding and coastal inundation by the rising seas. But these forecasts of sea-level rise are in concomitant decline. In 1980, then-federal climatologist Steven Schneider (now at Stanford) forecast sea levels rising tens of feet; he even published a map showing the Washington Monument awash in the tidal Potomac River in an outrageous paper that appeared in the 1980 peer-reviewed *Annual Review of Energy*. That the reviewers would approve the publication such a ridiculous forecast is a testimony to the times.

Those predictions were based upon early model scenarios that forecast an unrealistically large amount of warming to take place in the high latitudes, thereby melting the vast volumes of ice stored there. Since then, IPCC sea-level rise forecasts for the next hundred years have ebbed to the range one to three feet.

There are two primary reasons that sea levels rise: more water and "thermal expansion." We must examine the effect of each independently to assess the future impact of climate change. A rise in

159

global temperatures does affect each, but regionality and seasonality of the temperature change are important when assessing total overall impact.

Water sources that could potentially contribute to sea-level rise include surface and ground water, glaciers and ice caps, and the ice sheets of Antarctica and Greenland. Of the three, the one humans most directly affect is the alteration of surface and ground water storage. Water diversions, changes in stream networks, changes in watershed characteristics and irrigation practices all affect the terrestrial hydrologic cycle. No one has fully analyzed the sum of those changes and therefore the effects on sea levels are not well quantified. Rough forecasts from changing land and water use practices' contribution to sea-level rise are rather low, with the 1996 IPCC estimate at about 1.1 inches over the next 50 years. In any event, the changes are totally unrelated to the amount of carbon dioxide in the atmosphere. They have occurred and will continue to occur regardless of how much "extra" CO_2 human activities emit. Therefore, we should not consider sea-level rise due to direct alteration of the hydrologic cycle in discussions of the climate change aspects of sea-level rise.

Far more important are the possible contributions from the vast stores of ice on the planet. Nonpolar glaciers around the world have been melting during the past 150 years or so. The initial cause of the melting was the rebound of temperatures from the multicentury period of lower temperatures known as the Little Ice Age, which ended in the mid-19th century. Before 1850, nonpolar glaciers experienced rapid accumulation and growth of ice. Since then, however, temperatures have been rising, and the glaciers have been melting. The 1996 IPCC report estimates that the additional melting of these glaciers will contribute about 6.3 inches of sea-level rise by the year 2100. Considering that estimates of the total amount of water contained in nonpolar ice are in the range of 5.9 to 10.6 inches, as A. Ohmura and others noted in a 1996 *Journal of Climate* article, the IPCC estimate seems much too high, in that it would require a melting of 60 percent and 100 percent of all the world's ice outside of Greenland and Antarctica!

The fact is, the largest cause of glacial melting is a prolonged warming of summer daytime temperatures, but winter warming is the hallmark of the enhanced-greenhouse era, and winter temperatures have no effect on melting. It is during summer days that temperatures on the glaciers tend to reach above freezing.

By far, most of the world's ice is contained in the vast sheets over Antarctica and Greenland. If those regions were to release large volumes of water, drastic sea-level rise would indeed occur. But temperatures in those places are extremely low, with only the margins and southern portions of Greenland subject to any melting at all during the course of a year. Even future warming scenarios do not change that fact. With higher temperatures, though, more moisture is available in the atmosphere, and more snowfall occurs in those regions. The result is net snow and ice accumulation in the cold areas, which include nearly all of Antarctica and the interior portions of central and northern Greenland.

In fact, the 1996 Ohmura study found that, over the course of the next hundred years, the amount of water the ice on Antarctica contains is supposed to increase, producing a sea-level drop of 3.54 inches, while the ice sheets on Greenland experience a net reduction, adding about 4.33 inches of sea-level rise. The numbers from the 1996 IPCC report are in a similar range for these regions.

Therefore, using the future warming scenarios from current climate models—and including sea-level rise from the addition of new water to the oceans not due to land and water use changes—ocean-level rises are forecast to be between 6 and 8 inches (6 inches from midlatitude ice and 1 inch from Greenland/Antarctica) from the melting of land ice.

Another source of sea-level rise is thermal expansion. As the temperature of water increases, its density decreases, so that a constant mass of water occupies a larger volume at higher temperatures. In other words, water expands when it warms. Global warming during the past 100 years is thought to be responsible for about 1.5 inches of sea-level rise caused largely by thermal expansion, according to the 1996 IPCC report. The IPCC forecasts for the next 100 years call for an additional 11 inches from thermal expansion. Considering that global mean temperatures have risen about 0.6 degrees in the last hundred years and are forecast by the IPCC to rise a mean of 2.5° (4.5°) in the next hundred years, the IPCC estimate of sea-level rise due to thermal expansion is clearly too high. Based upon observations alone, the sea levels should rise only five inches by the year 2100 as a result of expanding sea waters.

In the most recent IPCC report, the "best guess" forecast for the total sea-level rise by the year 2100 is 19.3 inches (which includes land and water use changes). Another scenario included in the same IPCC report and described by the IPCC as "internally consistent, plausible, and 'state-of-the-art,'" shows only 10.2 inches of rise. Our argument here leads to an expected value of 10 inches, based upon the notion that the IPCC must be overestimating the amount of midlatitude melting and thermal expansion. Whatever the assumptions, the IPCC has lowered its own forecasts significantly from its original 1990 report, in which the median sea-level rise estimate to the year 2100 was 26 inches. The primary reason for this reduction was that midrange forecasts of global temperature rise dropped from 3.2°C to 2.5°C. But the new scenarios are based upon GCMs that increase the future concentration of atmospheric carbon dioxide faster than is currently forecast and much faster than has been observed, at least according to James Hansen in the 1998 *Proceedings of the National Academy of Sciences.* Other contributory findings, noted previously, include too much warming from methane (the rate of increase of which was overestimated) and an overestimate of the direct warming effects of carbon dioxide, as Myhre noted in 1998 in *Geophysical Research Letters.*

Taking into account these changes, the net total annual temperature rise in the next century reduces to an amount very close to the value we noted at the beginning of this chapter using the "linear model" argument of chapter 6.

Applying the reduction in temperature to the forecasts of sea-level rise results in a sea-level rise of between 50 percent and 60 percent of the current forecast values. Even using the IPCC's two "equally plausible" scenarios, the best guess range of sea-level rise during the next century would be about 5 to 11 inches—a rise that most people might well not notice and to which they could easily adapt. After all, much of the city of New Orleans is currently *below* sea level.

We can and do adapt to changes that are this slow. For example, largely because of geological (rather than climatic) considerations, relative sea level in Tidewater Virginia has risen one foot during this century. Few people are aware of this change, except the residents of Tangier Island, a barrier at the mouth of the Chesapeake Bay that— like all barrier islands—is dynamic and unstable. Tangier Island is slowly disappearing.

Sea-level rise does not make for disaster unless it occurs in very large amounts over a short time period. A change of that magnitude could occur only if the West Antarctic ice sheet collapsed. Though some scientists have proposed that a warmer world would greatly increase the likelihood that this will happen, research published by Charles Bentley in *Nature* in 1997 showed the chance of its occurring is about 1 in 100,000 in any given year. The current conformation of the ice sheet is quite stable, and there is no evidence of any stress that might indicate a sudden shift. Further, it takes thousands of years for the ice sheet to respond to changes in surface air temperatures because a major instability requires shifts in the flow of the ice streams at the bottom of the sheet, which is grounded on the sea bed and thus insensitive to surface temperatures.

Therefore, all indications are that few humans will be directly impacted by rising sea levels—whatever their cause. A possible exception? Residents of low-lying island nations. But they are already dealing with sea-level rise. In the Maldives, for example, where nearly 80 percent of the land has an elevation of less than three feet, residents are now taking steps to better protect their most important shorelines from the effect of storms and waves, regardless of global warming.

Sea-level rise has been characterized by the 1996 IPCC report "in terms of environmental and social consequences . . . [as] arguably one of the most important impacts of global climate change." But this is off base. The most likely scenario during the next century is one of a slow rise in sea level, not much different from the one that has occurred over the last 100 years. To this we have adapted, and we will continue to do so into the future.

El Niño

The strong 1997–98 El Niño generated historic levels of media coverage, making it easily the most discussed climate story in human history. That El Niño was repeatedly linked to nearly every severe weather event that occurred over that time period, including flooding in California, strong Atlantic coastal storms, tornadoes and wildfires in Florida, and drought in Texas, to name a few. And for the first time, some people connected El Niño to global warming.

Few of the links survive careful scientific scrutiny. Although our understanding of El Niño and its impacts improves with each El

Niño event, it is clear that we still know very little about direct causes and effects of this recurring climatological feature.

El Niño and its flip side, La Niña, are described in chapter 3. As a refresher, recall that El Niño is a weakening (or a reversal) of the Pacific Ocean trade winds and that it is measured by observing the difference in atmospheric pressure between Darwin, Australia, and Tahiti.

In *Geophysical Research Letters* in 1996, Trenberth and Hoar looked solely at the Darwin pressure data. They concluded there has been a relatively high prevalence of El Niño events in the last two decades. Using these observations, they claimed that, based on statistical probabilities, such a prevalence and strength could only occur statistically once every 2,000 years. In a rebuttal in the same journal, Harrison and Larkin challenged their conclusion, noting that Trenberth's data set was significantly smoothed and filtered, and their statistical analysis suggested that "it is plausible . . . that the unusual behavior in the early 1990s is the result of natural variability."

And in a modeling study using the NCAR GCM, published in *Geophysical Research Letters* in 1998, Liang and Wang found that, without any increases in greenhouse gases, prolonged and frequent El Niño events are not rare at all in portions of the record—another indication that the recent increase in El Niños is likely part of a naturally occurring cycle.

Later, during the height of the 1997–98 El Niño event, the media quizzed Trenberth about the global warming–El Niño connection. After first noting that the models we have are not really good enough to convincingly make this connection, Trenberth nevertheless wrote in the *UCAR Quarterly* that "there's got to be a connection." Administration officials, including the vice president, repeatedly used this general line of thinking, claiming that the weather events they linked to El Niño are typical of what we can expect as global warming becomes more pronounced. Echoing Trenberth, Gore proclaimed in a June 1998 press release (the same one in which he distorted the 1998 tornado frequency), "We know that as a result of global warming, there is more heat in the climate system, and it is heat that drives El Niño."

Several additional modeling studies that specifically examined El Niño events contradict Gore and Trenberth. Using the Princeton GCM, Tom Knutson and Sukuryo Manabe predicted a reduction in

El Niño activity in a future atmosphere with elevated carbon dioxide concentrations; their article appeared in *Geophysical Research Letters* in 1994. De-Zheng Sun, using a different modeling approach, likewise demonstrated an El Niño reduction as the planet warms in the same journal in 1997.

Because El Niños have been occurring for thousands of years, it is possible to extract information on them from the geologic record. As we noted in chapter 3, Sandweiss et al. detected almost no evidence of El Niño activity 4,000 to 7,000 years ago during a period when the planet was 1°C (1.8°) to 2°C (3.6°F) warmer than it is today. Grove, using historical texts, noted that the earth has a long history of severe El Niños and proclaimed in a 1998 *Nature* article, "There have been colossal El Niños [over many centuries] and there was no global warming then." Based on an oxygen isotopic analysis of corals, Fairbanks developed a several-hundred-year record of El Niño occurrence and found that current El Niño levels are similar to those of the late 18th century, in the midst of the Little Ice Age. In the 19th century, when it was warming as the Little Ice Age waned, El Niño occurrence declined.

A 1998 *Geophysical Research Letters* article reported somewhat contradictory results from Tsonis, who posed the possibility that El Niño/La Niña events have a role in *balancing* global temperature trends. Tsonis hypothesized that during warming trends, El Niños become more frequent and serve as a method of heat dissipation. In short, El Niños and La Niñas may be responsible for regulating the planet's temperature, making sure the planet gets neither significantly warmer nor colder.

Despite numerous proclamations to the contrary, well-established, predictable linkages between El Niño events and weather occurrences outside the tropics are rare. This is partly because each El Niño event is unique, and the resulting atmospheric circulation in the middle latitudes is highly variable. Hoerling and Kumar, in a 1997 *Geophysical Research Letters* article, compared upper air circulation for some recent major El Niños and found that "most of the observed differences in large-scale circulation anomalies among the [El Niño] events may not [be] predictable from knowledge of [El Niño]." The variability between El Niños is so great that some regions experience flooding during one event and drought during the next.

165

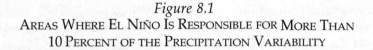

Figure 8.1
AREAS WHERE EL NIÑO IS RESPONSIBLE FOR MORE THAN
10 PERCENT OF THE PRECIPITATION VARIABILITY

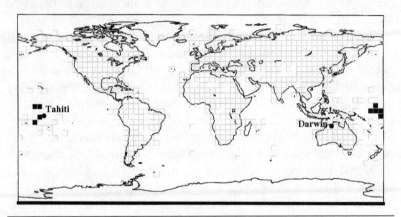

NOTE: This figure shows regions of the world where 10 percent or more of the precipitation variability is explained by the standard measure of El Niño. Given the propensity to blame every weather event on El Niño, the area is remarkably small.

Our map (Figure 8.1) shows the areas of the world where 10 percent or more of the variation in total rainfall is explained by changes in the standard measure of El Niño. Here, we correlated monthly values of the Darwin–Tahiti pressure difference to total monthly rainfall using the United Nations' precipitation record.

The highest correlation is in the tropical Pacific, which should not be very surprising, since this is where El Niño is. The correlation coefficient—a measure of the correspondence between El Niño and rainfall—reaches its largest absolute value in the western Pacific at –0.56. The negative sign means that El Niño conditions bring rainfall shortages to the region. Mathematically, the percentage of correspondence between two variables is the square of the correlation coefficient, which in this case works out to 31.6 percent.

Elsewhere, however, the measure of the extent to which El Niño corresponds to rainfall is much, much lower. Indeed the magnitude of the average correlation coefficient between El Niño and rainfall over the world is a paltry 0.07, or an explained variance of a whoppingly small 0.5 percent. Over North America, no correlation is

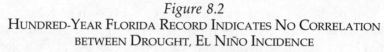

Figure 8.2
HUNDRED-YEAR FLORIDA RECORD INDICATES NO CORRELATION
BETWEEN DROUGHT, EL NIÑO INCIDENCE

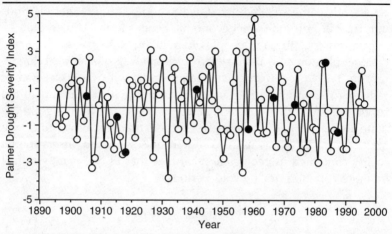

NOTE: This figure shows annual values in Florida of the Palmer Drought
Severity Index (PDSI), a commonly used indicator of precipitation/soil
moisture status, with solid circles representing El Niño events highlighted.

higher than 0.20, which results in an explained variance of only 5
percent. (Yet the tendency to link anything and everything climatic
to El Niño seemed to soar somewhere around 95 percent!)

The winter of 1997–98 was noteworthy in and of itself for the
impact of a number of strong Atlantic coastal storms, or nor'easters—
but of course many people claimed these were a response to the
contemporaneous El Niño event. Historical records of strong nor'-
easter occurrence along the mid-Atlantic, however, indicate that no
statistically significant relationship exists. Nor is there a relationship
between total nor'easter counts and El Niño.

The spate of wildfires along the central Florida coast in early
summer, driven by severe drought conditions, was frequently connected with El Niño, especially by Vice President Gore. But no
relationship exists between drought in this region and El Niño
events. Figure 8.2 shows Florida's Palmer Drought Severity Index
annual values, with El Niño events highlighted. El Niños are as
frequently related to wet years in central Florida (positive PDSI) as
to dry years.

One of the few demonstrable linkages is between El Niño and Atlantic hurricane activity—which El Niño suppresses. And that may be a good thing. Because tropical cyclones are one of the major causes of weather-related damage, it could be argued that El Niños may actually provide a net economic benefit to the United States.

In summary, although people often link El Niño events to specific weather phenomena outside the tropics, analysis of historical climate records fails to confirm such relationships, with the exception of reducing the number of Atlantic hurricanes, which originate in the subtropics. And though scant indications (mainly from certain climate models) suggest to some researchers that global warming will make future El Niños more common or severe, the observational reality contained in geological records is that El Niños may in fact be more common during cold periods.

9. Global Warming and Mortality

Will mortality increase with rising temperatures? Death rates during periods of extremely hot weather have jumped in certain cities, but above-normal mortality has not occurred during all hot spells or in all cities. Moreover, research concerned with "killer" heat waves has generally ignored or downplayed the reduction in fatalities that warmer winter months would bring.

The United States is one of few highly populated countries that maintains a long-term, high-quality database on mortality. For the latter half of the past century, everyone who dies has been entered into a master database that includes his or her age, race, sex, county of residence and death, date of death, and cause of death (from the death certificate). This database provides a potentially excellent source for evaluating environmental relationships to human mortality.

To address possible future weather–mortality relationships, we must first establish current linkages. Laurence Kalkstein and Robert Davis, in the *Annals of the American Association of Geographers* in 1989, prepared one of the most comprehensive studies based on daily mortality and weather data for all the major metropolitan areas in the United States. Their results highlight the importance of local factors in mortality, making the development of a generally applicable national or global model difficult, if not impossible. For example, Kalkstein and Davis determined that some cities have a specific "threshold temperature," a temperature beyond which mortality increases abruptly. But no straightforward, linear relationship exists between temperature and mortality counts. In Chicago, for instance, when 4 p.m. apparent temperatures are below about 30.5°C (87°F), daily death totals can be high or low (Figure 9.1). As temperatures reach the upper 80s and beyond, however, the likelihood of higher death totals increases.

But not all cities have threshold temperatures—hence the importance of intercity differences and acclimatization. For example, there

169

Figure 9.1
CHICAGO TOTAL DAILY MORTALITY VS. 4 P.M. TEMPERATURE

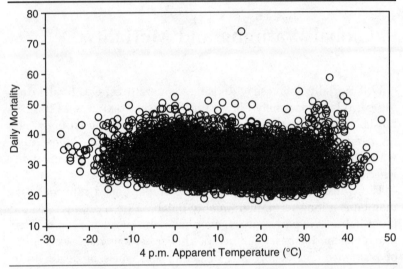

NOTE: Deaths increase sharply when temperatures rise above 30.5°C (87°F).

is no summer threshold temperature in Phoenix or Jacksonville. In those cities, no increase in mortality occurs whatever the maximum temperature. Populations there get accustomed to hot conditions, with the help of better building design, the prevalence of air conditioners, physiological adaptation, and other factors. Similarly, Minneapolis has no low threshold temperature in winter. Generally, though, most northern cities have summer threshold temperatures, and most southern cities have winter threshold temperatures, since the people in these regions have not adapted to extreme conditions. So a 34.5°C (94°F) high in Jacksonville might have no mortality impact, while similar conditions in Boston could produce high death totals.

So how might mortality rates in a city such as Dallas change in response to global warming? Simply for the sake of argument, let us assume that there will be more warm days in Dallas in the future. Will more people die because more days exceed the threshold temperature? Or will the threshold temperature become higher as people adapt? Or will the relationship of mortality to temperature change so that Dallas has a mortality–temperature situation more like what

Phoenix has now, with no threshold temperature? Those people promoting worst-case scenarios simply assume the current relationship would carry into the future, and that mortality would increase dramatically. But is this realistic?

To correctly forecast future mortality changes, we must consider the ongoing changes in our infrastructure. In the United States, for example, half of all homes were built within the last 25 years. No doubt the mass of them were fitted with central air conditioning, which is known to reduce summer mortality. Many more old buildings with poor ventilation have been upgraded with central air or window units, and these changes must be factored into future mortality scenarios.

Although high-quality mortality data are available in the United States, small sample sizes continue to limit city-specific research. Proper analyses are possible only for the largest metropolitan areas, but even those data sets may not be large enough. Daily mortality data have high variances, even in the largest cities. In Chicago, note that there are numerous low mortality counts on some of the hottest days and that the variance grows with increasing temperature (Figure 9.1). That kind of problem is magnified for cities with smaller populations.

Furthermore, most of the analyses focus on total mortality to provide the largest possible samples. In some cities, certain significant relationships for total mortality do not hold for the elderly, nonelderly, ethnic minority, or nonminority subsets. Relationships are often inconsistent across cities. As Thomas Gale Moore noted in his 1998 *Climate of Fear*, in some cities, women are most susceptible to high temperatures; in others, men are. In *Environmental Health Perspectives* in 1991, Kalkstein showed that hot weather significantly impacts blacks in St. Louis but not in New York City, where whites are disproportionately impacted. All of these relationships may be real and explainable, but they could also be statistical artifacts. Information on cause of death in the mortality data set provides little real insight. The reason is simple: Few Americans die specifically from heat or cold exposure; rather, during adverse weather conditions, heart attacks and strokes, two of the most common causes of death, occur slightly earlier than they would have otherwise. For example, in the major Chicago heat wave in July 1995, Cook County Medical Examiner Edmund Donoghue explained the victims were

171

probably very near death and that their date of death was merely moved up by the heat. "How long they would have lived, I can't tell you," coroner Barbara Richardson of nearby Lake County told the *Chicago Tribune*. "There's no way you're going to get me to say that definitely these were heat deaths. If it is 20 degrees below zero and someone dies of a heart attack, is that a cold death or a heart attack death?"

These officials' comments highlight two additional problems with mortality statistics. First, after a major heat wave, subsequent days typically have total deaths well *below* the average, which suggests the heat may only be advancing death in the susceptible populace by a few days. Second, even a professional medical examiner has trouble determining whether heat or cold was ultimately a significant factor in a given death.

Forecasts of increased future mortality from global warming must consider differences between winter and summer mortality and ways the future climate will change. In the United States, more people die in winter than in summer (Figure 9.2). Based on a 10-year record, there are an average of 132 heat deaths and 385 cold deaths annually, according to Moore. For almost all death categories, winter mortality exceeds that of summer. Overall winter mortality is 16 percent higher than that of summer, a statistic that is partially attributable to communicable diseases, such as influenza and pneumonia, that flourish during cold weather.

Almost all global warming theory predicts that most of the warming will occur in the high latitudes and in winter. Furthermore, most of the warming is occurring in the coldest air masses—the ones responsible for the winter cold air outbreaks. Warming of those air masses would presumably reduce future winter mortality rates. In comparison, if the observed trends of the last third of the 20th century are meaningful, summer warming will be about 60 percent of what occurs in the winter. When the additional future use of air conditioning is considered, summer mortality rates could very well decline, even *with* a modest warming. One counterbalance could be an energy tax, which government would presumably impose to mitigate global warming and reduce death rates; in fact, the opposite could result, as individuals opt to reduce their air conditioning use in hopes of saving money. Such a tax may very well ultimately increase mortality. Like the balance between lighter, more fuel-efficient automobiles

Figure 9.2
Comparison of U.S. Weather-Related Deaths

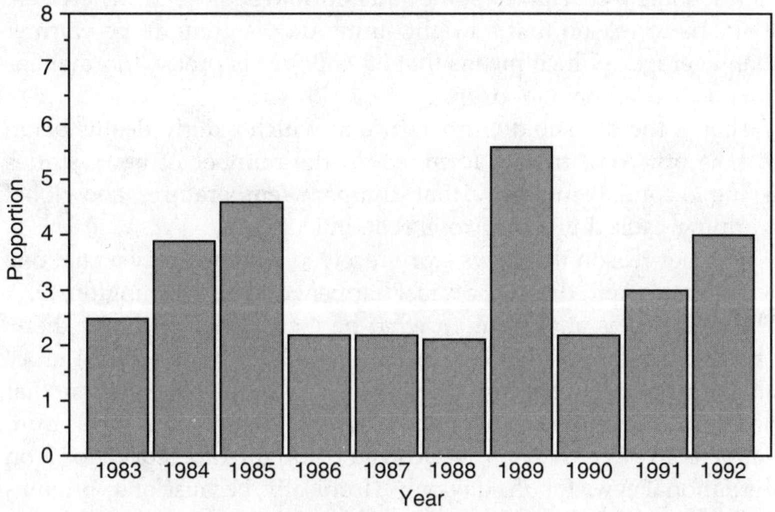

SOURCE: *Vital Statistics of the United States, 1983–92.*
NOTE: This figure shows the ratio of weather-related cold deaths to heat stress mortality.

and traffic deaths, tradeoffs and unexpected consequences inevitably arise from policy actions.

Heat-Related Mortality and Air Conditioning

The summer of 1999 was hot, but not unusually so. That fact did not stop the media from incessantly reporting the number of "heat-related" deaths, which came in at about 180. Most of the deaths occurred in the eastern United States.

How unusual were these circumstances? Does the enhanced greenhouse effect, and everything else that goes along with it, make the situation worse or better?

The average July high temperature along most of the East Coast from New York on down is around 32°C (90°F). Sometimes a sea breeze wafts into the Big Apple, but Philadelphia, Baltimore, and Washington are far enough inland that they simply bake.

Given that the last week of July is, climatologically, the warmest in that summer month, temperatures in the lower 90s (°F) should

173

be the rule, not the exception. But these "normal" values comprise 30-year averages. Some of the days that form that average were sunny, some were cloudy, some had rain, and some were in between. It stands to reason that a bright sunny day is going to be warmer than average—which means that 35°C (95°F) is pretty "normal" as long as it does not cloud up.

That is the threshold temperature at which elderly deaths begin to take off. And, true to form, so do the number of news stories trying to conflate mortality, that summer's temperatures, and global warming caused by pernicious economic activity.

Another reason this news story received so much play is that one of the hottest and driest spots just happened to be Washington, D.C. The situation is analogous to what happens when it snows there. The last big dump, two feet in January 1996, was the No. 1 story on the network news for days, and it was that snowstorm that prompted the infamous *Newsweek* cover. When such a snowstorm happens in say, Madison, Wisconsin, it might rate story No. 3 on the national news for one day only. (Ironically, because of its proximity to the Gulf Stream, 24-hour snow records are actually greater in D.C. than they are in Madison.) In these instances, perception carries more influence than reality; it seems hotter (or colder, or drier, or wetter, or more extreme), so it must be.

In fact, there is no warming trend in U.S. summer temperatures over the last 80 years (Figure 9.3). Temperatures did warm a bit from 1900 to 1930, but that surely was not a greenhouse effect change; after all, we had not put much new carbon dioxide into the air by then. Pre–greenhouse-effect 1930 is still the benchmark hot spell in the mid-Atlantic this century. And 1999 did not come close, in either length or severity. Further, summer 1999 planetary temperatures, as measured by satellites and weather balloons, were below their average for the last two decades.

What is more, heat-related mortality is going *down*. In 1995, Chicago saw several hundred deaths in a July heat wave. But back in 1955, there were 885 heat-related deaths in the Second City. Want to see true carnage? Go back to 1901, when 10,000 Americans perished in the heat. (The globe was 1°F cooler then!)

What is the difference between 1901 and 1999? Two words: *air conditioning*.

Air conditioners use more electricity than other home appliances. On a hot day, they create such demand for electricity that, sometimes,

Figure 9.3
U.S. SUMMER TEMPERATURE HISTORY

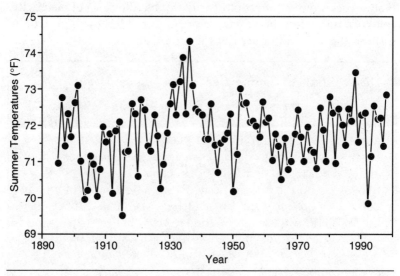

NOTE: Summer temperatures in the United States show no increase in the last 80 years.

the power fails. Once that happens, the county coroner is not far around the corner. In the end, it was a power failure that magnified the 1995 Chicago tragedy. Normally, in a heat wave, the poorer South Side—with less access to air conditioning and less income with which to pay for it—experiences more deaths than the North Side. But a power outage in the affluent side of town results in a pretty equal distribution of fatality across income classes.

This truth is self-evident: The very technology that enhances the greenhouse effect—the production of electricity—is what saves our lives in the heat of summer. Without air conditioning, thousands more would die, as they did in 1901 in a world where the enhanced greenhouse effect and dreaded global warming did not even exist.

Of course, we can avert the risk of power failure by installing new generation capacity. Yet every time a new power plant is proposed, someone squawks "global warming." When lack of power causes an outage on a hot day, that well-intended protest becomes a lethal weapon.

Therefore it is somewhat ironic that all proposals to fight global warming drastically raise the price of energy and power. The UN Framework Convention could easily double the price of electricity. If it goes into effect, how many more of our elderly will hesitate to turn on the air conditioner until it is too late?

Direct Effects of Carbon Dioxide on Human Health

At today's level, around 365 ppm, carbon dioxide poses no human health threat whatsoever; commercial greenhouse growers find 1,000 ppm to be a plant-pleasing standard they can live with day to day. But at high concentrations, those above about 15,000 ppm, CO_2 does begin to have notable negative effects on the human body. Excessive amounts of blood-borne CO_2 produce acidosis, which can be fatal. But it is just not possible for this concentration to occur in the earth's atmosphere as a result of the combustion of fossil fuel; we cannot burn fuel fast enough to emit enough CO_2 to achieve this concentration.

Concentrations typical of the last 10,000 years range from 260 to 320 ppm. As technological civilization has risen, so have concentrations, which are currently about 365 ppm. Assuming an average residence time for a molecule of carbon dioxide of 100 years (estimates range from 13 to 200 years, depending upon the "sink" in which the molecule is captured), human beings simply do not have the foreseeable industrial capacity to bring atmospheric concentrations anywhere near the toxic level. The 1996 IPCC report estimates that, with the maximum possible exponential increase in emissions and deforestation, concentrations still would be 15 times lower than the toxic threshold by the year 2100. Of course, even these attempts to predict technological behavior at the hundred-year time horizon are highly speculative; the assumption that our energy system will even remain fossil fuel–based is an open question, given the technological history of the last three centuries.

10. Greening the Planet

Recent findings about climate change and plant physiology suggest that the notion of planetary ecocide from global warming is simply wrong, and that its opposite—a greening of the earth—is occurring.

In this context, the observation that warming predominates in the cold winter air masses in Siberia and northwestern North America is particularly relevant. The growing season (the time between the last spring freeze and first fall freeze) usually begins and ends under high-pressure systems that form in these regions. If those systems warm, then the air masses that are cold enough to define the growing season appear earlier in the spring and later in the fall. Is there any evidence that this change is occurring?

Writing in *Nature* in 1997, Myneni and four coauthors published a stunning, if controversial, finding. Using a rather brief satellite record beginning in 1981, they found that the high latitudes were "greening up" about a week earlier by the 1990s. In commentary, NASA scientist Compton Tucker noted that these latitudes usually have very short growing seasons, and as a result, "We're talking about a 10 percent increase in carbon uptake by plants." Everything else being equal, then, high-latitude and arctic vegetation is about 10 percent larger because of a possibly human lengthening of the growing season.

Myneni's finding corroborated an obtuse study published by Thompson of the Bell Laboratories in *Science* in 1995, which found that the spring warmup had progressed about three days forward in the Northern Hemisphere's midlatitudes.

In October 1998, S. Fan and several colleagues published a bombshell article in *Science* that found that North American vegetation is growing so fast that it may be absorbing the *entire* annual carbon dioxide emissions from the United States and Canada! They attributed this to four factors, the last of which is a warming of the high latitudes, which serves to lengthen the growing season, increasing

177

THE SATANIC GASES

the amount of time for plants to absorb carbon dioxide. They cite three other factors. One was regrowth of forests on U.S. lands previously devoted to logging and agriculture. Urban sprawlers—who decry the minuscule loss of farmland (compared with the nation's total) that gets eaten up by cities—might beware here. In fact so much farmland is being abandoned to forests that it is cited as a principal reason for the massive greening of North America.

Another factor Fan et al. cited is the effect of nitrogen fertilization from agriculture and industry. Here, Fan et al. are in agreement with Finnish forester Pekka Kauppi, who found in 1992 that Europe's forests—contrary to Greenpeace propaganda—are growing by leaps and bounds, thanks, among other things, to nitrogen fertilization.

The remaining reason, according to Fan, is the fertilization effect carbon dioxide has on plants. An overwhelming body of evidence shows the rising levels of atmospheric CO_2 are most favorable for the production of food and forests. CO_2 acts as a fertilizer, increasing plant growth rate and mass by increasing photosynthetic capacity. It increases plant water use efficiency and drought tolerance as well as performance under low light conditions and in high temperatures. Carbon dioxide also increases plants' abilities to grow in the presence of environmental hazards imposed by soil alkalinity, mineral stress, atmospheric pollutants, and UV-B radiation.

Such facts are painstakingly documented in Sylvan Wittwer's 1995 book, *Food, Climate and Carbon Dioxide*. Wittwer served as chairman of the National Research Council's Board on Agriculture, and his book is the single most comprehensive review of this subject. In his final section, Wittwer concludes:

> The debates on global warming and its magnitude will likely continue without resolution well into the 21st century. . . . The effects of an enriched CO_2 atmosphere on crop productivity, in large measure, are positive, leaving little doubt as to the benefits for global food security. . . . Now, after more than a century, and with the confirmation of thousands of scientific reports, CO_2 gives the most remarkable response of all nutrients in plant bulk, is usually in short supply, and is nearly always limiting for photosynthesis. . . . The rising level of atmospheric CO_2 is a universally free premium, gaining in magnitude with time, on which we can all reckon for the foreseeable future.

178

Wittwer discusses hundreds of experiments that have confirmed that rising levels of atmospheric CO_2 have enhanced plant growth, total plant output, and the yields of all the major food crops (cereals, legumes, roots and tubers, sugar crops, fruits, and vegetables). In his estimation, global agricultural output has increased 8 percent to 12 percent in the last 50 years due solely to the rising levels of atmospheric CO_2.

Precisely *how* carbon dioxide makes things greener is a story worth detailing. Much of this discussion can be found in Wittwer's writings; what follows here merely scratches the surface.

Enhanced Photosynthetic Capacity

As everyone knows, plants, using the energy of light, remove carbon dioxide from the air and synthesize it into plant fiber in a process called photosynthesis. Plant respiration is the process by which plants consume this plant material to maintain life. Enhancing atmospheric CO_2 levels increases the efficiency of photosynthesis and markedly reduces plant respiration. Using less energy in the respiration process means that more energy can be allocated to growing larger. As far as agricultural crops are concerned, this increased efficiency results in increases in total dry weight, root growth, higher root–top ratios, leaf area, weight per unit area, leaf thickness, stem height, branching and seed, and fruit number and weight. This improvement makes for an increase in marketable product as well as an overall shortening of maturing time—reducing both water and pesticide requirements and expanding geographic growth range.

There are three types of plants: C3, C4, and Crassulaceous Acid Metabolism (CAM), categorized by the type of photosynthetic pathway they employ for CO_2 use. Each of these plant types responds differently to CO_2 increases; some researchers have hypothesized a competitive advantage from increased CO_2 levels might arise that favors one of these groups over another. The University of Michigan's James Teeri, for instance, has championed the notion that as the CO_2 content of the air continues to rise, C3 plants (which comprise about 95 percent of all plant species and include most major crops except corn and sugar cane) will outcompete C4 plants (which comprise about 1 percent of plants, including corn, sugar cane, sorghum, millet, and some grasses), thereby driving many of them to extinction. Teeri and others base their theory on the fact that C4 plants typically exhibit less of a CO_2-induced growth stimulation than C3 plants. (CAM plants, by the way, are mainly desert succulents that are agriculturally insignificant.)

In 1993, H. Poorter surveyed the literature in this area in a review article published in the journal *Vegetation*. Poorter found that C3 plants, on average, exhibited a 41 percent increase in growth, while C4 plants displayed a 22 percent stimulation, for a doubling of the air's CO_2 content. But conditions in these studies varied widely and, when the conditions are taken into account, the differences between C3 and C4 responses to enhanced carbon dioxide become much smaller.

Many scientists, including us, have predicted that as the atmospheric CO_2 content continues to climb, earth's mean air temperature will also. The amount and distribution is a subject of serious scientific contention (hence this book). But, given that almost all scientists believe that enhancing CO_2 will create some warming, how might the phenomenon influence the relative distributions of C3 and C4 plants?

Because they are so diverse, C3 plants can be found in nearly all environments on earth, whereas C4 plants are typically relegated to hot and/or arid climates, such as those associated with tropical and desert regions, where they thrive as a consequence of unique biochemical and anatomical adaptations they possess. One of the ways C4 plants have adapted is in their ability to increase the CO_2 concentration at the point at which it turns into plant carbohydrate, which makes C4 plants more efficient at gaining carbon dioxide. The adaptation greatly reduces carbon losses from respiration directly associated with photosynthesis, which can account for the "cannibalization" of up to 50 percent of recently stored carbon in C3 plants, as Ivan Zelitch showed in *Biological Science* in 1992.

Although this CO_2 concentrating mechanism requires additional cellular energy to operate, it does provide C4 plants with a competitive advantage in environments with high air temperatures, such as what scientists are predicting for earth's future. In areas with cooler temperatures, however, C3 carbon losses decline, and the additional energy required by C4 photosynthesis no longer provides C4 plants with an advantage over C3 species. More than 20 years ago, in the journal *Oecologia*, Teeri and Stowe (1976) documented this temperature-driven phenomenon in a study of C4 grass distribution across North America. They determined that the percentage of C4 grasses in local flora decreased with increasing latitude, actually going to zero for all locations within the Arctic Circle. They also determined that C4 grass distribution in the continental United States was more strongly correlated with July daily minimum temperature than with any of the other 18 environmental variables they considered. Consequently, Teeri and Stowe concluded that "the warmer the nights, the greater the success of C4 taxa."

Because nighttime temperatures in July have the greatest influence on C4 plant distribution, a little nocturnal warming could well stimulate their poleward expansion; and studies of land-based records have shown that most of the 0.6°C (1.1°F) warming observed in the last 100 years has occurred at night. Consequently, as CO_2-induced warming proceeds, C4 plants should be able to persist in their current locations and may even expand into regions where they do not now exist, given the low night temperatures that occur there.

In thousands of experiments, the data show that elevated CO_2 levels favor C3 plants and elevated temperatures to favor C4 plants. Consequently, since the planet of the future will experience both higher CO_2 levels and higher temperatures (especially at night, the most crucial time for C4 plants), it is hard to say which of these phenomena will have greater influence on the competition between C3 and C4 plants. But the compensatory effects of CO_2 and night warming may suggest little relative change.

Will Climate Change Be Too Fast for Plants?

Some people have expressed concern that global warming driven by elevated levels of atmospheric CO_2 will be so great that plants will need to migrate toward the poles or up mountainsides to remain within the climatic regimes to which they are currently adapted. If climate change is too rapid, they fret, plants will not be able to migrate fast enough to avoid extinction. Seeds do not have feet and plants cannot run.

But basic plant physiological research largely contradicts such concerns. In a comprehensive analysis of 42 different experiments, appearing in 1994 in *Agricultural and Forest Meteorology*, Keith Idso and Sherwood Idso found that the percent growth enhancement

181

resulting from a 300 ppm increase in the air's CO_2 content actually rose with increasing air temperature going from close to zero at 10°C (50°F) to 100 percent at 38°C (100°F). This increase in relative growth response arises from the fact that the growth-retarding process called photorespiration is most pronounced at high temperatures but is effectively inhibited by atmospheric CO_2 enrichment. So powerful is this effect of elevated CO_2, in fact, that the optimum temperature for plant growth and development typically rises with increasing CO_2 levels. Dozens of researchers in plant physiology have duplicated this result.

One of the first researchers to quantify this change in the form of an hypothesis this was S. P. Long, in a 1991 article in the journal *Plant, Cell & Environment*. Using well-established plant physiological principles, Long calculated that the optimum growth temperatures of most C3 plants should rise by approximately 5°C (9°F) for a 300 ppm increase in the air's CO_2 content. And in a voluminous number of scientific studies that have experimentally evaluated this phenomenon, a 300 ppm increase in atmospheric CO_2 has been found to cause several C3 plants' optimum plant growth temperatures to rise by about 6°C (11°F). It is also worth noting that the photosynthetic rates of these particular C3 plants were found to be nearly twice as great at their CO_2-enriched optimum temperatures as at their optimum temperatures under ambient CO_2 concentrations. Consequently, not only would the predicted increases in atmospheric CO_2 and air temperatures not hurt earth's vegetation, but they would likely work synergistically to promote its growth and development, as ever more investigations demonstrate.

We state with confidence that doubling the air's CO_2 content causes a rise in optimum C3 plant growth temperature that is even larger than is predicted to result from co-occurring CO_2-induced global warming. Clearly, such warming would not adversely affect the vast majority of plants; for fully 95 percent of them are of the C3 variety. The remainder of the planet's species are largely tropical and will not experience quite so large a CO_2-induced rise in optimum temperature, but then they are already adapted to the earth's warmer climates and use the other photosynthetic pathways (C4 and CAM). Bear in mind that the tropics are expected to warm much less than other portions of the globe, according to both observed data and IPCC.

Even at the highest air temperatures plants encounter, atmospheric CO_2 enrichment is still desirable. Indeed, it can mean the difference between life and death. Countless studies show that elevated CO_2 typically enables plants to maintain positive carbon exchange rates in situations (such as very hot environments) where plants would otherwise exhibit negative growth rates that ultimately led to their demise. This type of research fills such journals as *Environmental and Experimental Botany* and the *American Journal of Botany*.

It is abundantly obvious from a large body of detailed scientific investigations that a CO_2-induced warming would not produce a massive poleward or up-slope migration of plants seeking cooler weather. After all, the temperatures at which nearly all plants perform at their optimum would rise even higher than the temperatures of their respective environments under such conditions. Indeed, elevated levels of atmospheric CO_2 will enable most plants not merely to cope with predicted air temperature increases, but to *thrive* in their presence, performing even better than they do today.

"Plant Cooperation" and Carbon Dioxide Enhancement

In 1997, S. W. Simard, writing in *Nature*, demonstrated that trees pass nutrients along a complex network of fungal matter with an abundance of nutrients at its disposal to whatever is lacking them, regardless of species. This rather stunning finding suggests that competition among plants may not play as great a role in natural ecosystems as scientists in the past believed they did. Instead, as this nutrient-sharing phenomenon would appear to discover, species coexistence and greater biodiversity are a biological imperative. Enhancing root systems improves this capability.

It is also well documented, by researchers such as J. S. King, S. A. Prior, P. S. Curtis, and dozens of others in many journals, that elevated levels of atmospheric CO_2 enhance below-ground growth and stimulate the root activities of most plants. One such CO_2-enhanced process is the exudation of nutrients and carbon compounds, which stimulates microbial and fungal activities in the vicinity of plant roots. As the air's CO_2 content continues to rise, it should lead to the development of ever better networks for distributing nutrients among plants, enhancing their transfer from the "haves" to the "have-nots," including both C3 and C4 species. This observation calls the whole concept of "competition" into question and suggests that "cooperation" may be a more fitting term to describe interspecies interactions in a future world of higher CO_2.

Considered in their entirety, these observations provide no substantive basis for believing that C3 plants will outcompete C4 plants and drive any large portion of them to extinction as the air's CO_2 content continues to rise. If anything, they point to the tantalizing possibility that both types of plants will fare even better in the future than they do now, and that they may actually help each other to some degree, given the way that opportunities for cooperation among species arise more often with elevated CO_2-induced increasing root growth and fungal networking below ground.

Increased Water-Use Efficiency

Another essential plant phenomenon that rising levels of atmospheric CO_2 improves is water-use efficiency. Water stress is the single greatest factor limiting global food production. By increasing plants' ability to make best use of the available moisture supply, CO_2 directly enhances plant growth.

CO_2 improves water use efficiency by closing down the pores (stomates) through which plants lose moisture. In their review article, Idso and Idso found that when plants received inadequate water, the percent growth enhancement from CO_2 became even greater than for well-watered plants. The relative enhancement for well-watered plants was 31 percent; for moisture-stressed plants, 62 percent, a twofold increase. Plants also increase their fine root mass—which increases their ability to ingest water—at higher CO_2 concentrations, according to research published in *Ecology* by P. S. Curtis in 1990.

Again, increased water-use efficiency means plants growing in higher temperatures (which increases moisture stress by increasing evaporation) should also experience a relative growth enhancement as carbon dioxide increases.

Other Environmental Stressors

In 1998, A. C. Volin, writing in *New Phytologist*, described the interactive effects of elevated CO_2 and ozone (O_3) on the growth of C3 and C4 species. In general, at elevated ozone levels, plants growing at ambient CO_2 manifested lower growth and photosynthetic rates than those growing in reduced ozone concentrations. But the deleterious

184

effects of high ozone concentrations were reduced, and in some cases even eliminated, when plants were exposed to elevated levels of CO_2. This amelioration occurred irrespective of plant photosynthetic pathway or growth form, demonstrating that both C3 and C4 plants should better withstand exposure to this and other forms of air pollution as the atmospheric CO_2 concentration continues to increase.

Other air pollutants, such as nitrogen and sulfur oxides, have known negative effects on plant growth as well. Elevated CO_2-related percentage increases in plant growth were generally greater in the presence of those substances. The mechanism may be similar to the one that increases water efficiency: When the stomates close, they restrict the pollutants' access to sensitive plant tissues.

Low light levels are another factor elevated carbon dioxide does more than just guard against. Doubling the concentration results in more than a doubling of low-light plant growth, compared with concentrations that existed in the atmosphere before the combustion of fossil fuels.

The Historical Record of Crop Yields

It is clear from an overwhelmingly large number of scientific studies that rising atmospheric CO_2 levels are a net benefit to agriculture and crop production. Figure 10.1 shows the combined global yield for major foods crops for the last half-century as well as the global temperatures. It indicates that regardless of the temperature, food production has increased steadily.

In the United States, the picture is much the same. Figure 10.2 shows the historical record of soybean yields in Illinois, a large producer of this crop, and of the temperature history there during the past three-quarters of a century. Figure 10.3 shows the same for corn in Iowa; Figure 10.4, for wheat in Kansas.

Everywhere, the picture is the same. Yields continue to increase, primarily as a result of technology, but undoubtedly elevated atmospheric CO_2 levels have helped as well. Global temperature rise and global yields for combined corn, soybeans, and wheat—the major crops—*are rising*. Iowa and Kansas temperatures stay the same, and yields of corn and soybeans *rise*. Illinois temperatures *fall*, and soybean yields *rise*.

Figure 10.1
A 50-YEAR RECORD OF CROP INCREASES

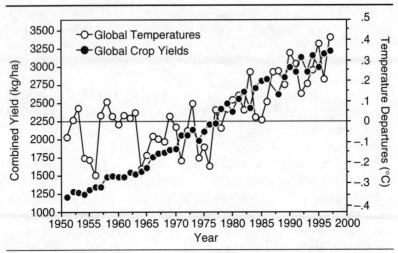

NOTE: Over the past 50 years, global crop yields have increased steadily, whether the temperature has been warming or not.

Figure 10.2
75-YEAR HISTORY OF SOYBEAN YIELD INCREASE IN ILLINOIS

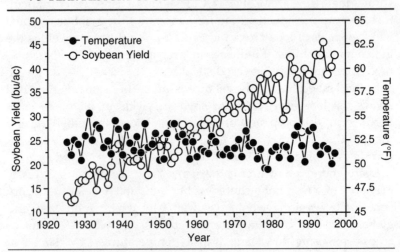

NOTE: Soybean yields in Illinois (open circles) have been steadily rising over the past 75 years, while temperatures (solid circles) there have declined.

Figure 10.3
75-Year History of Corn Yield Increase in Iowa

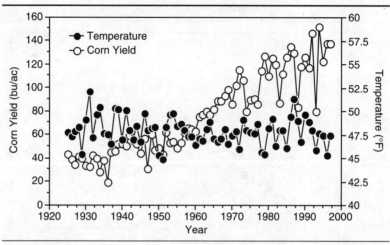

NOTE: Corn yields in Iowa (open circles) have been steadily rising over the past 75 years, while temperatures (solid circles) exhibit no significant change.

Figure 10.4
75-Year History of Rising Wheat Yields in Kansas

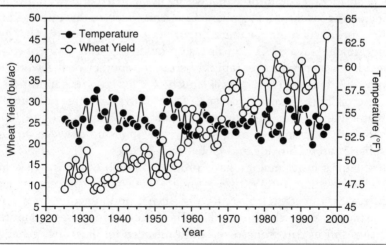

NOTE: Wheat yields in Kansas (open circles) have been steadily rising over the past 75 years, while temperatures (solid circles) there show no significant change.

187

Figure 10.5
THE POSITIVE STATISTICAL RELATIONSHIP BETWEEN CO_2 AND
GLOBAL CROP YIELDS

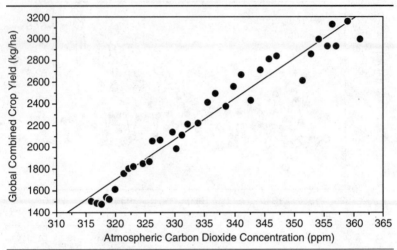

NOTE: A strong statistical relationship exists between global crop yields and atmospheric carbon dioxide concentration.

Consider it good fortune that we are living in a world of gradually increasing levels of atmospheric CO_2. The effects of this increase on food production are far more important than any putative change in climate. Elevated CO_2 levels also provide a cost-free way of conserving water, which is rapidly becoming another of the world's most limited natural resources, much of it for crop irrigation.

According to Sylvan Wittwer, CO_2 is a "universally free premium" that enhances crop growth without additional inputs. Figure 10.5 shows the extremely strong statistical relationship between atmospheric carbon dioxide and global crop yields.

Changes in technology have prompted much of the increase in yields over time, but Wittwer estimates that some 10 percent of that increase is directly related to CO_2—which is undoubtedly true given the nature of the change in yield. The temporal increases in Figure 10.6 result in part from increases in nitrogen fertilizer use, genetic and mechanical improvements, and tillage practices. But none of those variables is changing in such a smooth fashion over time. The *only* agricultural input that changes so smoothly is carbon dioxide.

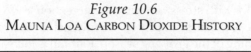

Figure 10.6
MAUNA LOA CARBON DIOXIDE HISTORY

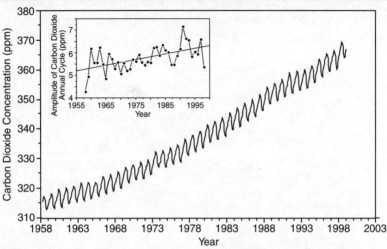

NOTE: The atmospheric concentration of carbon dioxide shows an annual cycle due to plants' uptake of carbon dioxide during the Northern Hemisphere summer. Inset: The amplitude of the annual cycle is increasing, an indication that plants are taking up more carbon dioxide—which is to say, they are growing better.

For further indirect evidence that CO_2 enriches agricultural yields, look at Figure 10.6, which depicts CO_2 concentrations measured at Mauna Loa since records began in 1958. The data's annual cycle is the planetary greening that takes place every year as a result of the dominance of vegetation in the Northern Hemisphere. Inset in Figure 10.6 is the history of the amplitude of that cycle. Clearly, as CO_2 increases in the atmosphere, an increasing greening of the natural vegetation results. Agricultural plants behave no differently from others with respect to their CO_2 uptake—they *must* be experiencing the same annual growth enhancement that the rest of the world's vegetation is.

The story that emerges from the world's vegetation is one of remarkable internal consistency. Carbon dioxide makes plants grow better—whether the temperature stays the same or rises, whether plants encounter stress or not, and especially when it rises coincidentally with stress. Altering the greenhouse effect causes climate change to occur primarily in the cold half of the year, resulting in a lengthening of the growing season. Rainfall is increasing slightly. All of these circumstances can only make the planet greener.

11. How Did We Get Here?

No one has successfully deconvolved the interrelation between science and society. But everyone who has considered the problem recognizes its complexity. The international nature of the global warming adds yet another layer to the puzzle.

In all highly developed economies, science is a government activity. There is insufficient gain for individuals or specific industries to provide the massive amounts of funding comprehensive studies of issues such as global climate change require. Whether that is good or even desirable is a reasonable subject for debate, but, as a society, we now feel that such undertakings are the province of government.

Public servants recognize this fact and attempt to use it for what they deem is a perceived good. For decades, one of the great supporters of science was the late Rep. George Brown (D-Calif.), who was well aware of the potential importance of climate change. He held hearings on climate change nearly 25 years ago, when the planet had just completed a cooling phase and a series of agrometeorological and agroeconomic disasters had resulted in a low point in world grain storage. In the 1980s and '90s, Rep. and then Sen. Albert Gore chose a similar path, this time emphasizing the perils of global *warming* rather than global cooling. His 1992 book, *Earth in the Balance,* articulates the beliefs he developed in that process. Gore's model is that science should precede policy. When government funds science, however, the research process becomes inextricably mired in the political one.

It is unavoidable. For example, on February 26, 1992, Gore chaired a hearing of the Commerce Committee's Subcommittee on Science, Technology, and Space, to discuss NASA's budget. This particular subcommittee oversees not only NASA's budget but also those of other major science agencies such as the National Science Foundation.

At the time, NASA wanted a major budget increase for something called the Earth Observing System with which to monitor climate and atmospheric chemistry. Midway through the hearing, Gore said he believed that the study of global warming should be NASA's highest priority. "With regard to the study of climate change," Gore said, "I think this is NASA's No. 1 priority now, or ought to be. And I think it is *our Nation's No. 1 scientific priority"* (italics added).

In the past 15 years, the total federal outlay on global climate change research has ballooned from a few million dollars to $2.1 billion per year. The proposed budget for Fiscal Year 2000 raises the figure to $4.1 billion. Much of the increase began in Gore's subcommittee; since Gore moved on to the vice presidency he has maintained a keen interest in its activities.

It is in this political environment that our nation's science administrators compete for funding. So do all scientists. This level of expenditure makes any scientific problem a serious political concern that is likely to develop into a sociopolitical issue. For example, *Earth in the Balance* was published just two months after the February 1992 hearing. It is impossible to imagine agency administrators encouraging their employees to speak out and bite the hand that feeds them.

This potentially places science and scientists in an uncomfortable paradox. Every scientist I know believes he or she is pursuing some form of objective truth. That is the lure of the business. Most federally funded scientists truly believe that their work is more "pure" than that which comes from industry because there is no obvious financial imperative to please stockholders. They consider the public funding process to be more "value free" because it is filtered through institutional reviews by peers and superiors.

But that seems naïve. George Mason University's James Buchanan won a Nobel Prize with his much different point of view, known as "Public Choice Theory." Simply put, Buchanan argues that individuals in the public service—scientists, administrators, technicians, and the like—put their own self-interest first. When a monopoly source of funds appears (our federal government, for example, is certainly very close to being the sole provider of research funding for climate change), and that source is biased toward one political view or another, then the recipients of the funds will support that political view, Buchanan argues.

In *The Calculus of Consent*, the 1962 book in which Buchanan (with Gordon Tullock) laid the groundwork for Public Choice Theory, he wrote, "We must assume that individuals will, on the average, choose 'more' rather than 'less' when confronted with the opportunity for choice in a political process, with 'more' and 'less' being defined in terms of measurable economic position."

Speaking to the nature of individuals vs. groups, Buchanan and Tullock write, "Under individualistic postulates, group decisions represent outcomes for certain agreed-upon rules for choice after the separate individual choices are fed into the process."

In this way individuals build political support for whatever produces personal gain. Not exactly shocking, but certainly counter to the popular perception of value-free scientists in search of objective truth.

This cause-and-effect relationship between issues and investigation is real. Science requires a lot of money and, in the public sphere, large amounts of dollars do not appear without political support. So an issue *must* become politicized for major research funding to materialize.

Public Choice Theory and the Theory of Scientific Revolutions

Perhaps the most influential treatise ever published on the nature of science is Thomas Kuhn's *The Structure of Scientific Revolutions*. Published the same year (1962) as *The Calculus of Consent*, its core ideas join with those of Buchanan to synergistically provide a model that may explain much of the evolution of the notion of disastrous global warming. Taken together, Kuhn and Buchanan predict that large amounts of research support will generate a "paradigm," or an overarching theoretical structure that is believed by scientists to explain the majority of a system's behavior. Kuhn calls the activity of most scientists "normal science," which is basically the care and feeding of the paradigm, as well as its defense, suppressing the publication or the importance of novelties that assault the paradigm.

As quoted, Kuhn wrote, "Mopping-up operations are what engage most scientists throughout their careers. . . . Closely examined, whether historically or in the contemporary laboratory, that enterprise seems an attempt to force nature into the preformed and relatively inflexible box that the paradigm supplies."

Employing several examples from physics and astronomy, Kuhn notes that the response of most scientists is often to resort to increasingly ornate and bizarre explanations when the paradigm is threatened by inconvenient realities. "Its defenders will do what we have

193

already seen scientists doing when confronted by anomaly. They will devise numerous articulations and *ad hoc* modifications of their theory in order to eliminate any apparent conflict." Thus the "models" of the earth-centered universe became increasingly complicated in an attempt to keep the faith. The paradigm of the climate models was assaulted by a lack of predicted warming, so the models were modified with other compounds—sulfate aerosols—to counter the warming. As we saw, this "*ad hoc* modification" does not work.

The belief that the currently conceptualized computer simulation global warming is accurate is just another paradigm. One interesting aspect of Kuhn's theory is that paradigms are defined by the communication process, largely through scientific journals. Scientists who edit the journals are not demonstrably different from others; that is, they are largely interested in maintaining the paradigm. Ironically, it is those editors who largely determine what the holy writ of the paradigm becomes!

Do they have preconceived agendas? Steve Schneider, now of Stanford University, told *Discover* Magazine in 1989 that he has to choose "the right balance between being effective and being honest" about global warming. For three decades, he has been the chief editor of *Climatic Change*, the first scientific journal devoted to the study of climate and its social implications.

Kuhn argues that the likelihood a paradigm will fall is very low unless there is something to replace it. In other words, you cannot just go around saying that the climate models were wrong; there must be a rational explanation. In the case of rapid and disastrous global warming, the IPCC recognized the paradigm failure in 1996 when it wrote that the greenhouse-only models "predicted more warming than has been observed," and it also offered, in the same sentence, a counter-paradigm: "unless a lower sensitivity" of the climate to greenhouse gases is used. It offered another alternative: that sulfate aerosols are responsible for the dearth of large warming.

This book argues that it is the sensitivity, not the sulfates, that was the problem.

Having said that induces our own dilemma. How can we defend this hypothesis without resorting to the same scientific literature we are claiming is likely to show a general bias?

A Kuhn/Buchanan perspective has its limitations—otherwise we would rarely encounter something we like to call truth. Instead, what we find is that a smaller number of papers and findings that are a part of that literature in themselves provide a counter to the existing paradigm. If we did not have these, there would be no *Satanic Gases.*

In other words, the Kuhn/Buchanan perspective provides only a general explanation for the tendencies that evolve because of the interaction between the political and scientific processes. This perspective does not provide an adequate explanation of the scientific ego, which contains a heavy intellectual counterweight: the desire for discovery and recognition as an individual and not as part of the paradigm. Kuhn and Buchanan predict systemic rather than individual behavior.

Nonetheless, the academic reward structure is biased heavily toward the Public Choice model Kuhn modified.

A university spends its every research dollar under the authority of a senior scientist, usually referred to as a principal investigator (PI). PIs must have academic appointments, and they are usually at senior levels (associate or full professor) at universities. These appointments are either "tenured" (jobs in perpetuity) or "permanent," with advancement and honors largely predicated by research productivity. In federal laboratories PIs enjoy analogous reputation and demands, with the exception of teaching.

Taxpayers have doled out around $8 billion in global climate change research money in the last four years. It is a pretty fair bet that there are fewer than 1,000 people nationwide who could qualify as PIs to carve up this pie. A fair rule of thumb is that about half of the outlay goes for hardware and overhead. That still leaves about $4 million per PI, on the average, over the period. Some programs, especially modeling programs at federal laboratories, consume much more than that. Others, such as data analysis projects at universities, might chew through $600,000 over the three-year period. Spread over five years, even the lowest-level PIs spend about $1.25 million. High rollers exceed $10 million.

Building a Scientific Paradigm via the Reward Structure

These billions of dollars buy a tremendous amount of research because, by and large, academic labor is in large supply and the wages are low. Very few PIs do very much actual research; instead they leave it to their junior colleagues, whom they direct. After all,

195

it is difficult to run a large climate model and write grant proposals at the same time. So the work goes mainly to the postdoctoral students (waiting for the PIs to retire, when they can take their place on the academic ladder), and the lowly graduate students. Neither of these two is paid very much. Graduate student research assistantships average around $12,000 per year, and postdoctoral researchers feel fortunate to receive $40,000.

This army of thousands of underpaid, high-IQ foot soldiers desperately wants academic positions that can lead to tenure or permanence, and the competition is severe. At major universities, beginning assistant professorships, which are seven-year probationary appointments, attract hundreds of applicants per job offered. The appeal of job security is evident in the pay: assistant professors, eligible for tenure or permanence, average around $45,000 per year, just a bit more than the postdoctoral fellows.

Every major university now summarily throws out applications from graduate and postdoctoral students aspiring to assistant professorship who have not published in the refereed literature before receiving their doctorates.

Scientists send papers to journal editors, who in turn send them out for review by other scientists. The process is not, as most people imagine, usually anonymous, at least in many of the atmospheric science journals. Instead, articles are sent out with the authors' names quite prominent, and many journals now allow authors to specify just whom they *do not* want to review the document.

The chance that a finishing graduate student in climatology owes his publications, his dissertation, and therefore his newfound job, to federal global climate change funding is very high. Who among them is going to write a dissertation that says global warming is an overblown problem? What editor is going to accept the paper without vigorous review?

And on the off chance that our hypothetical graduate student becomes an assistant professor, the rush to publication accelerates even further. Although it may take three refereed publications over the average five-year period between receiving a B.S. and a Ph.D. simply to land an assistant professorship, it takes a remarkable 30 publications or so at top-drawer universities for promotion to tenure or permanence. Most universities will not appoint a person beyond six years without this contractual change, so the evaluation process usually takes place in the candidate's fifth year.

In other words, an assistant professor has to generate six refereed publications per year to make it. That research requires money. And if the chosen career is climatology, there is only one real provider: the federal government.

During this process, the up-and-comer must pay obeisance to older, more established figures in his field because it is they whom will ultimately be solicited to write letters to the university in support of promotion. One lukewarm letter spells unemployment. One demigod in the field, irritated by a young upstart, can destroy a career with a few keystrokes.

The candidate must garner an appointment to a committee in his professional organization, such as the Applied Climatology Committee of the American Meteorological Society. That area of expertise, defined by publications and committee service, will encourage the editors of the big refereed journals such as *Nature*, *Climate Change*, and the *Journal of Climate* to send manuscripts for the young peer's review.

In this environment, manuscripts contending that global warming may not be all it is feared to be meet with especially vigorous review. Not a bad thing at all, in and of itself: Such strenuous perusal typically results in "counterparadigm" papers that are very high quality, if relatively infrequent. The cloud in this publish-or-perish sky is that such articles are more difficult to publish. But the silver lining is that those papers that do make it into the literature have proven absolutely compelling to the journal editor and reviewers.

Indeed, despite the Public Choice pressure, a sufficient number of persuasive studies that paint a radically different picture from the doomsday forecasts have met with such scrutiny and made it into print. This book alone contains references to, or uses information from, hundreds of refereed articles in support of its own "counterparadigm," including portions of the IPCC reports.

Eventually, in the long run, the truth emerges. But in the short run, money buys paradigms, and paradigms drive policy.

We now come to the ticklish question of the role of the vice president in promoting global warming gloom and doom. Was Al Gore just a passive observer who happened to oversee the major science funding on this issue, or is there evidence of a broader environmental agenda?

Recall that he stated that global climate change should be NASA's "No. 1 scientific priority." On the face of it, that is not a particularly bad sentiment; after all, everyone in a position of authority must have his priorities. What is disturbing are the tactics to which his writings allude. In *Earth in the Balance*, Gore states his unwavering position. Is this not his current agenda also?

> We are now engaged in an epic battle to right the balance of our earth, and the tide of this battle will turn only when the majority of people in the world become sufficiently aroused by a shared sense of urgent danger to joint an all-out effort. It is time to come to terms with exactly how this can be accomplished [p. 269].

And on the next page,

> Though it has never yet been accomplished on a global scale, the establishment of a single shared goal as the *central organizing principle for every institution in society* [italics added] has been realized by free nations several times in modern history [p. 270].

Finally,

> Adopting a central organizing principle—one agreed to voluntarily—means embarking on an all-out effort to use every policy and program, *every law and institution, every treaty and alliance* [italics added], every tactic and strategy, every plan and course of action—to use, in short, every means to halt the destruction of the environment and preserve and nurture our ecological system. Minor shifts in policy, marginal adjustments in ongoing programs, moderate improvements in laws and regulations, rhetoric offered in lieu of genuine change—these are all forms of appeasement, designed to satisfy the public's desire to believe that sacrifice, struggle, and a wrenching transformation of society will not be necessary [p. 274].

Without question, Gore has followed through, and it is easy to see which "programs," "institutions," and "laws" were his concern. Perhaps most important, though, is the "treaty," which is obviously the Framework Convention on Climate Change, signed in 1992. Gore has been quite faithful to his own script.

But paradigms have consequences. The next chapter discusses the policy responses.

12. Treaties, Programs, and Protocols

The Framework Convention on Climate Change and the Kyoto Protocol

In 1992, the IPCC produced a Supplementary Report "to provide the technical basis ... for a 'Framework Convention on Climate Change'" (FCCC), addressing what it called "key conclusions and new issues." It was this report negotiators turned to at the Rio de Janeiro United Nations Conference on Environment and Development, known more commonly as the "Rio Earth Summit," in June of that year.

The treaty's objective is to "stabilize greenhouse gas concentrations in the atmosphere at a level that [prevents] dangerous anthropogenic interference in the climate system." Signed by President Bush and soon approved by the U.S. Senate, it went into effect March 21, 1994, six months after 50 nations ratified it. The Framework Convention on Climate Change is a treaty with great potential to dictate the domestic energy policy of its signatories. As such, it represents a potential major transfer of national sovereignty to an international authority.

Are we at the "dangerous" level of interference, the level *Merriam Webster's Collegiate Dictionary* calls "able or likely to inflict injury"? If anyone can demonstrate that enhancing the greenhouse effect causes any injury whatsoever, must we reduce emissions to the level beneath which that injury will not occur? Is that a "dangerous" level of interference? Or must the net effect be injurious—which is to say that the beneficial effects do not balance out the bad ones?

If that is the case, stabilizing greenhouse gases at current concentrations requires us to reduce our emissions by 60 percent to 80 percent. In the United States, meeting that standard would depress our net use of fossil fuels to somewhere around the 1930 level.

Curiously, some of the most rapidly growing greenhouse gases— gases the U.S. Environmental Protection Agency had said would be responsible for nearly 25 percent of prospective global warming if left unchecked—were omitted from the Framework Convention. These are the chlorofluorocarbons, used mainly in refrigeration and covered under a separate UN treaty on stratospheric ozone depletion. The Framework Convention's intent was obvious: By disallowing credits for CFC reductions, the United Nations confirmed that large net reductions in *any* greenhouse gas emissions were insufficient. Their foremost mission was to curb fossil fuel use.

Less than two years after the Framework Convention on Climate Change was signed, it became apparent that global emissions of greenhouse gases were continuing to climb rapidly. In the United States, between 1990 and 1998, carbon dioxide emissions had risen a considerable 10.4 percent, according to the U.S. Department of Energy's highly reputable Energy Information Administration. If you assume a merely linear trend (which is a change from the century's previous behavior, when emissions rose exponentially), we can expect a rise between 1990 and 2010 of 23 percent. If you assume the exponential rate that is more common, then the increase by 2010 will be 37 percent. The observed value seems likely to fall between these two assumptions.

"COP-1" and the Berlin Mandate

The nations that signed the Framework Convention agreed to meet again after Rio, in March 1995 in Berlin. The first meeting of the so-called "Conference of the Parties" ("COP-1") was designed to iron out some of the difficulties inherent in the FCCC, in particular, the relative commitments of developed vs. poorer countries. The result was the "Berlin Mandate," which specifically exempted India, China, Mexico, and other poorer nations. It turns out roughly 80 percent of the world's population has no commitments under the FCCC.

Rhetoric in Berlin was certainly hotter than the weather. "An unseasonal cold snap . . . prevented plans for a 40-foot-high pyramid of ice at the conference convention center," the *Los Angeles Times* reported. "Its melting would have illustrated global warming, but some feared it would not have melted."

Opening the conference, German Environment Minister Angela Merkel said, "The greenhouse effect is capable of destroying humanity." Vice President Gore accused anyone who disagreed with his global-warming-as-catastrophe viewpoint of being "intellectually, politically, and morally bankrupt."

The problem confronting the Berlin meeting was that the FCCC was not "legally binding." It merely expressed a "goal" to reduce emissions to 1990 levels by the year 2000, which is a far cry from a legal commitment. As a result, the Berlin meeting hammered out an agreement that so-called "targets and timetables" would be agreed upon at the COP-3 meeting to be held in Kyoto in December 1997. What was more important—and what ultimately served to condemn the subsequent Kyoto Protocol to death in the U.S. Senate—was that the Berlin conferees agreed that developing and poor nations would be exempt from any specific commitments to limit greenhouse gases. That egregious gaffe, which only nations substantially ignorant of the way U.S. public opinion works could have made, would ultimately result in a unanimous Senate resolution expressing major reservations about proposed changes to the FCCC.

"COP-2": Targets and Timetables

The next meeting ("COP-2") was in Geneva in July 1996. Its purpose was to prepare the groundwork for hard emissions reductions "targets and timetables" that would replace the original convention's loose "goal." This ratcheting up of policy required equivalent attention to the apparent probity of global warming science. At COP-2, Under Secretary of State Timothy Wirth had said, "The science is convincing." He further stated that the IPCC had done its job:

> Our deliberations have benefited from the careful, comprehensive, and uncompromised work of the Intergovernmental Panel on Climate Change. Their efforts serve as the foundation for international concern and their clear warnings about current trends are the basis for the sense of urgency within my government.

Anyone who disagreed with the IPCC's conclusions, he attacked:

> We are not swayed by and strongly object to the recent allegations about the integrity of the IPCC's conclusions. These allegations were raised not by the scientists involved in the IPCC . . . but rather by naysayers and special interests bent on belittling, attacking, and obfuscating climate change science.

John Gummer, the British environment minister, saw ulterior motives in anyone who would disagree, including ourselves:

> None of us should give way to the commercial propositions which are hidden by the pseudo-science of those who pretend that what the world knows to be true can be put to one side because of an individual's desire to promote his particular and prejudiced view.

What "one individual" (Michaels) had done was to lay what appears in this book as Figure 6.3 on the literature table in the back of the delegates' meeting hall, where several nations that opposed the convention immediately seized it. Then the United States signed a Ministerial Declaration that agreed to the notion of "legally binding" emission reduction targets after 2000. Of course, the targets remained undefined, awaiting the COP-3 meeting in Kyoto, Japan, in December 1997.

The political process in the United States had awakened to what was happening. For one thing, much of the world was being exempted from any commitments to greenhouse gas reductions. And most econometric simulations showed that the treaty's potential costs were enormous for those nations that were obligated to it, even though specific reductions had yet to attach themselves to it. As a warning shot, On June 12, 1997, the Senate passed the Byrd-Hagel Resolution (S. Res. 98), by a 95–0 vote, stating that the United States should not sign

> any protocol or other agreement regarding the United Nations Framework Convention on Climate Change of 1992, at negotiations in Kyoto in December 1997, or thereafter, which would mandate new commitments to limit or reduce greenhouse gas emissions for the Annex 1 [Developed Nation] parties unless [it] . . . mandates new specific scheduled commitments to limit or reduce greenhouse gas emissions for Developing Country Parties within the same compliance period, or *would result in serious harm to the economy of the United States* [emphasis added].

The White House pressed forward, seemingly oblivious to the sense of the Senate. President Clinton, addressing the "Rio + 5" (a celebration of the fifth anniversary of the Earth Summit) meeting at the United Nations on July 7, 1997, gave a speech of remarkable hyperbole. He said that sea level will rise "two feet or more" in the next century. Recall that, at the time, the IPCC's median estimate was 19.3 inches and its alternative scenario of just 10.6 inches, was, in their words, "internally consistent, plausible and 'state of the art.'" Clinton went on to say that human-induced climate change would "disrupt agriculture, cause severe droughts, floods, and the spread of infectious disease." One year later, in 1998—an El Niño year and the hottest one ever measured in modern times—the world would set yield records for most major crops.

"COP-3"—The Kyoto Protocol

On October 22, 1997, President Clinton announced that the United States would go to Kyoto and agree to reduce our greenhouse gas emissions to 1990 levels by the year 2012. (He actually said "between 2008 and 2012," but that means 2012. Suppose the Environmental Protection Agency mandated that cars achieve a certain costly emission reduction by "between 2008 and 2012." Which car company would spend the extra money and raise their price in 2008?)

Clinton's proposal was too moderate for the Europeans, whose radically green environment ministers such as Merkel dominated the discussion. They wanted a 10 percent cut in emissions below 1990 levels. No doubt, their economic ministers had informed their governments that any severe reductions would impact the U.S. economy more than theirs, simply because energy is so readily and cheaply available here. Germany's structural unemployment of more than 10 percent was finally hitting home, and Merkel's proposal would certainly "solve" that problem by internationalizing it.

Back in the United States, Vice President Gore stroked moderates by saying U.S. negotiators would "walk away" from any bad deal. But anyone who had believed him has clearly not read *Earth in the Balance*. What people fail to realize is that his concept of a "bad deal" is closer to what Clinton proposed! With the conference deadlocked between the Europeans and U.S. proposals, Gore jetted in on the final scheduled days to instruct the U.S. negotiators to be more "flexible."

Some people say this was entirely planned—that Gore/Clinton agreed to put out a position so weak that the conference would deadlock and therefore require Gore's intervention. Whether or not that is true, "flexible" sounded to the U.S. negotiators like "give away the store." Most seasoned observers were agog when the

United States agreed to reduce its emissions to 7 percent below 1990 levels by the averaging period 2008–12. This works out to a stunning reduction in U.S. emissions of between 30 and 43 percent if we continued on our merry economic way, depending upon assumptions. These are the main points of the Kyoto Protocol:

- It is "legally binding" upon the signatories, allowing the United Nations to invoke whatever penalties it might choose upon those who do not meet their commitments.
- It commits the United States to a 7 percent reduction below 1990 levels in net greenhouse gas emissions over the period 2008–2012. It commits European Union nations and Canada to an 8 percent reduction. Australia is allowed an 8 percent increase.
- It commits none of the poor or developing nations, including China, India, and Mexico, to any emission reductions.

The economic costs have been estimated under various assumptions, which are very important and will need to be discussed before estimating the specific impact of Kyoto. They represent increases in the cost of energy mandated by the United Nations and the federal government in an attempt to discourage the use of fossil fuels to the point that we meet our commitments under the Kyoto Protocol. The amount required varies under different scenarios for how Kyoto will be implemented. The smallest in costs occurs with the use of "international emissions trading," which is the only feasible method for large reductions in net U.S. emissions.

Where Will This Come From?

The nation's greenhouse gas emissions are divided in roughly equal thirds among manufacturing, transportation, and electrical generation. Transportation shows little promise for any large net reduction in emissions. Population increases expected over the next 15 years guarantee an increased demand for automobiles that will be difficult to balance with changes in fuel efficiency. Given the time for new technology to diffuse into production of automobiles, it seems highly unlikely that vehicles manufactured 10 years from now will be radically different from those on the lots today. And we can expect that vehicles purchased today, including the plethora

204

of popular (if less than efficient) sport utility vehicles, are likely to be still on the road should Kyoto come into force in 2008.

Manufacturing relies upon much of the same infrastructure that individual consumers do; reductions in net output (i.e., recessions) are politically unacceptable. Rather, it is the infrastructure—mainly in electrical generation—where we could obtain the largest greenhouse emission reductions.

Currently, 56 percent of all American electricity is produced by the lowest-priced form of fossil fuel: coal. Coal produces about 30 percent more carbon dioxide per unit of energy obtained than natural gas (depending upon a number of variable assumptions about production and electrical efficiency). If the natural gas is burned in state-of-the-art turbines, compared with the currently on-line production, the net decrease in carbon dioxide emissions per unit power over coal is closer to 40 percent. Thus switching all coal-fired electrical generation to natural does not get us even close to meeting Kyoto's mandate.

That will not happen. The national infrastructure does not exist to supply this amount of natural gas. There is no known plan to create, in the few years that lie between now and 2008, a supply network to meet this demand.

The economic costs of Kyoto have been estimated under various assumptions, which are important and merit discussion before we estimate the specific impact of Kyoto. The amount of cost varies under different scenarios for how Kyoto is implemented. The largest reductions in cost occur because of something called "international emissions trading."

"Emissions trading" is an attempt to allow "market" forces to determine the price of compliance with the Kyoto Protocol. The assumption is that markets are far more efficient than command-and-control government intervention in dealing with costs of regulation.

Emissions trading creates emissions "credits" that nations can buy and sell among one another. First, each nation creates a national inventory of its greenhouse emissions, including the amounts from each specific economic sector and each corporation. Then, each corporation is required to submit an annual update.

Assume, for example, that the Russians are able to accomplish this task and that they plant a lot of trees, which sequester carbon dioxide. The amount of "saved" greenhouse gas gives them a "credit" that they can sell to another country. The price is determined by mutual agreement. If very few credits are available and demand for them is high, then so is the price. Consequently, Russia might be rewarded handsomely by the more affluent United States, which might choose to purchase its credit rather than substantially reduce emissions within its own boundaries.

The Department of Energy's EIA calculates that compliance with the Kyoto Protocol will cost the United States a stunning reduction in gross domestic product averaging 3.2 percent per year for the period 2008–2012, assuming domestic actions only and no emissions trading. A similar model by WEFA, Inc. also estimates a cost of 3.2 percent.

With complete emissions trading between industrialized, developed countries, the annual GDP change in an econometric model from DRI-McGraw-Hill associates becomes –1.6 percent, and in the EIA model, the reduction goes down to 1.0 percent.

The administration's Council of Economic Advisers (a more political and appointed body than the EIA) produced a remarkably low cost estimate of only 0.01 percent per year. In doing so, the council assumed complete and unfettered emissions trading between "Key Developing Countries" such as China, India, and South Korea.

Which of These Scenarios Is Likely to Be Correct?

Note that emissions trading first requires that an accurate inventory be made of national emissions from virtually every home and business. Then the inventory must be subject to international verification, presumably by a committee the United Nations would appoint (there seems to be no other alternative). Then the emissions reductions must be certified as genuine, a seemingly impossible task.

The fact is that, for most nations, we do not even know today how much carbon dioxide is actually emitted by all sources and subsequently sequestered by the biota with any large degree of confidence. We simply know too little about how carbon molecules move through the system, as all calculations based upon accepted parameters come up with the wrong answer, concluding that the concentration of carbon dioxide in the atmosphere should be much higher than it is. Either we are emitting less than we think or the biota are taking up more than we think, or some combination thereof.

It is fair to say that the balance of recent scientific evidence indicates that the credit largely lies in the biosphere, but no one knows for sure.

There are international political obstacles to emissions trading. The Kyoto Protocol may allow for emissions trading, but it does not provide a mechanism. That technicality was to be worked out at "COP-4," held in Buenos Aires in November 1998. But no substantive progress was made there.

One main reason was that many of the developed, industrialized nations, particularly in Europe, do not really want emissions trading. If the more affluent United States "buys" its way out of substantial domestic emission reductions, then less affluent nations cannot afford the increasingly expensive emissions credits. So European nations fear they will have to suffer not only disproportionately large reductions in emissions, but also economic disadvantage with the United States. Whether that is reasonable or not is debatable, but some people have suggested a compromise in which the proportion of emissions that can be traded not be allowed to exceed 50 percent. The EIA has analyzed this scenario and arrived at an annual GDP change of –1.7 percent.

In summary, a legally defensible emissions trading scheme is a few years away. So is 2008, the first year in which we are to somehow show massive reductions. We rapidly approach that date with no marketable scheme. The assumption that all of the developed industrialized countries will trade all their emissions is wrong. The administration's belief that all those countries, plus many developing nations, will do so, is a misleading fantasy. WEFA and EIA, assuming no trading, are much closer to reality than anyone else. The Kyoto Protocol will cost U.S. citizens a great deal of money.

Kyoto enjoys very little political support. Immediately after Kyoto, the *Washington Post* calculated a total of—maybe—12 votes for the protocol in the Senate, which requires a two-thirds majority for ratification. Rep. John Dingell (D-Mich.), regarded by friend and foe alike as one of the most perceptive and savvy American politicians of the 20th century, has said that the Protocol is "so flawed, in fact, that it cannot be salvaged."

Kyoto: All Pain, No Gain

Aside from its political liabilities, the Kyoto Protocol's fatal problem is that it will do nothing about global warming. No less a greenhouse luminary than Tom Wigley has calculated that, if all of the nations met all their commitments under the Protocol, that the amount of "saved" warming to the year 2050 would be 0.07°C (0.13°F). Given the year-to-year "noise" in lower atmospheric temperature data, which randomly swings around 0.10°C (0.18°F), this change in temperature will be truly impossible to detect. In agreement with our assertion about the general linear behavior of climate models, Wigley further calculated that reduction to the year 2100 that results from Kyoto is in fact twice this number, or a grand total of 0.14°C (0.25°F).

Further, the reductions Kyoto might induce will be distributed across the seasons in the same fashion as the warming it is supposed to mitigate, with the change in summer temperature considerably less than in winter. Here is a sobering thought: 20th century temperatures rose about 10 times the amount that Kyoto would prevent in the next 50 years, and, at the same time, life span doubled, crop yields quintupled, and the greatest democratization of wealth in the world's history took place.

13. The Future

We have no legal instrument that can effectively compel the community of nations to reduce greenhouse gas emissions enough even to reduce prospective warming by 10 percent. So what will the future bring? Many authors would now feel disposed to describe the world 100 years from now. We cannot do that because the only thing we really know about 100 years from now is that technology is likely to be so different from today's that it is virtually unpredictable. Think about 1900 vs. 2000. One hundred years ago no one knew what an atom was (much less an "electron," which powers this "computer," whatever that is), or what an airplane was. A world of rockets, thermonuclear weapons, and satellites? Unimaginable. People would have argued that today's median life expectancy of 80 years would be impossible; it was about 45 years in America in 1900.

Or compare 1900 to 1800. What in the world, would someone from 1800 ask, is a reaper? What is a railroad? Is the Mississippi River connected to the Pacific Ocean? The folly of technological prediction gives equally preposterous results, no matter what centuries you compare, once freedom looses human imagination. It is the height of folly to try to picture 2100.

Even predicting the next 50 years is a stretch, even on the well-studied subject of climate change. But you can expect the following:

1. The Kyoto Protocol will make no difference to the climate. Despite its enormous cost, the Kyoto Protocol will not appreciably change global temperatures from the trajectories that have been established in the last third of this century. It will not pass the U.S. Senate, and there will be no worldwide enforcement of mandatory reductions in greenhouse gas emissions that will be detectable as a change in climate from what would have occurred without regulation.

2. Carbon dioxide emissions will continue to increase. That has been the path of the world for the last third of a century. Developed economies, such as the United States', have become increasingly efficient. We now produce a unit of constant-dollar gross domestic product with about 60 percent of the energy that we used in 1970. But the rest of the world is growing, and is increasingly covetous of U.S. levels of comfort and convenience, so the net contributions of atmospheric carbon dioxide from developing nations more than balance the increases in efficiency taking place in the well-established world. The result? Carbon dioxide concentrations will continue to increase.

No one knows the exact rate of increase for carbon dioxide to 2050 because no one can predict what technological developments will arise to displace or augment fossil fuels. Perhaps the only thing we can say with any certainty is that there is no off-the-shelf substitute currently available that can compete economically. Because it takes so much time for substitute technologies to displace current ones (dictated in part by the length of time required to pay off financial obligations for new power plants and sport utility vehicles), we are unlikely to see much change in at least the next 25 years. Concentrations of atmospheric carbon dioxide will undoubtedly rise in the next 50 years.

3. Scientists will confirm that although the functional form of the climate models is correct, the amount of warming is already dictated by nature. We choose to believe that we have not completely wasted billions and billions of tax dollars on climate models, and that at least they have properly calculated that, despite exponential increases in carbon dioxide, the climate warms in a linear fashion. In surface temperatures for the last third of this century, a trend has already been established, and it is near or below the low limit of the model calculations, but it is a straight line, as the models predicted. There is no reason to suspect this is going to suddenly stop.

Further, the warming of the winter half-year is about twice that of the summer, and if the United States is any guide, precipitation has increased slightly. These trends will continue, but it is hard to see why they are deleterious. The warming of the summer is so small that it does not eat up the rainfall increase via evaporation.

4. The earth's average surface temperature will warm 0.65°–0.75°C (1.17°–1.35°F) by 2050. The distribution should be 0.75°–0.85°C (1.35°–1.52°F) in the winter half-year and 0.60°–0.65°C (1.08°–1.17°F) in summer. Scaling United Nations' figures on sea-level rise for these values gives an estimate of 3.0 inches over current values by 2050 using one of their models, and 5.3 inches using the other one, which they call "equally plausible." All these predictions take into account both the linear models and observed reality.

5. *As a result of increases in carbon dioxide alone, crop yields by 2050 will have risen by enough that the rise alone would feed one-quarter of today's world.* Assuming that Sylvan Wittwer is correct in ascribing a 10 percent increase in crop yield to current concentrations, it is easy to see another 10 percent to 15 percent by 2050. That adds up to a 25 percent increase.

6. *On a population-adjusted basis, temperature-related mortality will decline.* The number of cold-related deaths will fall. Because winter warms more than summer, cold-related deaths will decline at a faster rate than heat-related deaths will rise, if they rise at all. Expansion of the use of air conditioning, especially in poor countries, will further militate against heat deaths—so long as we get out of the way of letting their economies develop. Judging from historical records, the number of lives air conditioning saves in the United States alone each year is easily in the thousands.

Is There a Way Out?

How do we prevent another repetition of this Kuhn/Buchanan loop tape when the next global environmental "crisis" appears? Kuhn tells us to expect what we have already seen. Most scientists avoid rocking the dais, preferring the approval of their peers as they work to verify the existing paradigm. This strategy brings federal funding, and funding brings publications in the scientific literature, largely reviewed by fellow dais-steadiers. Buchanan tells us that individuals who are rewarded by the amount of federal funding they receive have an automatic incentive to behave in ways that enhance their chance of advancement.

Scientific paradigms are largely determined by the existing body of refereed literature. Scientists review scientists, and the approval process dynamic is the same for the literature as it is for continued employment at a university, where the activity of Kuhn's "normal science" is the surest route to a lifetime contract.

This seems like an unbreakable cycle. But it is not. Despite the claims that there are but a few "skeptics" promoting the views expressed in *The Satanic Gases*, there is a slew of studies that made

211

it into the refereed literature despite their defiance of the gloom-and-doom paradigm. Again, those papers must be of exceedingly high quality to have survived the review process in this environment. The truth is not completely suppressed. It just takes longer to get out than it should.

Society cannot afford to wait for the prevailing paradigm to change if it is to avoid making tragic policy blunders such as the Kyoto Protocol. Only in a Public Choice–driven climate of fear could such a document ever have emerged.

How can the self-corrective process of science be enhanced to prevent such tragedies? We must diversify the Public Choice bias inherent in any funding. Lower the federal outlay for research while increasing the flow from the private sector, including the industry and environmental communities. Perhaps federal research budgets in specific areas should be limited. Instead, additional funding can come from outside government, in lieu of corporate tax; a proportional mix of federal, environmental, and industrial oversight could then allocate those resources.

Of course, that would do nothing to alter the influence of Public Choice upon scientists. But it would widen the number of choices a scientist could make. The combined presence of the industrial and the federal viewpoints, along with that of the environmental community, could create a much broader "bias base" for scientists to serve. This would hasten the turnover of paradigms and accelerate publication of more diverse science more quickly in the refereed literature. In the end, a much more diverse group of university scientists would be rewarded with lifetime contracts—which is good for the educational system and good for the society it serves. In higher education, we now bludgeon science students with four years of what are de facto federal employees (who just happen to be professors). This is a disservice to our graduates, some of whom may actually want to work in private industry.

This proposal is more "back to the future" than anything new. In the 1997 book *Science for the 21st Century*, William Niskanen writes that private finance was the largest supporter of scientific research until the 1950s. With the launch of *Sputnik* and the subsequent orbital flight of Yuri Gagarin in 1961, the federal outlay to science rapidly outstripped that of private industry. Indeed, as Niskanen writes, "If the government funds much of the same type of activities as private firms, the government funds may reduce private support." Between 1955 and 1965, the federal outlay for science research more than *septupled*.

Replacing federal dominance to broaden the bias base sounds simple and logical, as simple and logical as science should be: the careful testing of theories and computer models with observed data. If we break the Public Choice stranglehold on scientific diversity, maybe science will return once more to that ideal. If we had done so earlier, maybe we would have had no need to clear the air about global warming.

References

Angel, J. R., and S. A. Isard. 1998. The frequency and intensity of Great Lake cyclones, *J. Climate*, 11, 1861–71.

Angell, J. K. 1994. Global, hemispheric, and zonal temperature anomalies derived from rawinsonde records. In *Trends 93: A Compendium of data on global climate change.* Washington, D.C.: U.S. Department of Energy, 984 pp. (and updates).

Angell, J. K. 1998. Contraction of the 300 mb north circumpolar vortex during 1963–1997, and its movement into the Eastern Hemisphere, *J. Geophys. Res.*, 103, 25, 887–93.

Angell, J. K., and J. Korshover. 1985. Surface temperature changes following the six major volcanic episodes between 1780 and 1980. *J. Cli. Appl. Met.* 24, 937–51.

Arrhenius, S. 1896. On the influence of carbonic acid in the air upon the temperature of the ground. *Philosophical Transactions* 41, 237–76.

Balling, R. C., Jr. 1992. *The Heated Debate: Greenhouse Predictions Versus Climate Reality.* San Francisco: Pacific Research Institute for Public Policy, 195 pp.

Balling, R. C., Jr. 1998. Analysis of daily and monthly spatial variance components in historical temperature records. *Phys. Geog.* 18, 544–552.

Balling, R. C., Jr., and S. B. Idso. 1989. Historical temperature trends in the United States and the effect of urban population growth. *J. Geophys. Res.* 94, 3359–63.

———. 1990. 100 years of global warming? *Envi. Consv.* 17, 165.

Balling, R. C., Jr., P. J. Michaels, and P. C. Knappenberger. 1998. Analysis of winter and summer warming rates in gridded temperature time series. *Clim. Res.* 9, 175–81.

Beersma, J. J., et al. 1997. An analysis of extra-tropical storms in the North Atlantic region as simulated in a control and 2 × CO_2 time-slice experiment with a high-resolution atmospheric model. *Tellus* 49A, 347–61.

Bengtsson, L., M. Botzet, and M. Esch. 1995. Hurricane type vortices in a general circulation model. *Tellus* 47A, 175–96.

———. 1996. Will greenhouse gas-induced warming over the next 50 years lead to a higher frequency and greater intensity of hurricanes? *Tellus* 48A, 57–73.

Bengtsson, L., E. Roechner, and M. Stendel. 1998. *Why is the greenhouse warming proceeding much slower than expected?* Max-Plank Institute Report No. 256.

Bentley, C. R. 1997. Rapid sea-level rise soon from West Antarctic ice sheet collapse. *Nature* 275, 1077–78.

Bodansky, D. 1994. Prologue to the climate change convention. In *Negotiating Climate Change: The Inside Story of the Rio Convention*, I.M. Mintzer and J.A. Leonard, eds. New York: Cambridge University Press, 45–76.

Bonan, G. B. 1998. The land surface climatology of the NCAR Land Surface Model coupled to the NCAR Community Climate Model. *J. Climate* 11, 1307–26.

Bonan, G.B., F.S. Chapin III, S.L. Thompson. 1995. Boreal forest and tundra ecosystems as components of the climate system. *Climatic Change* 29, 145–167.

Bove, M. C., D. F. Zierden, and J. J. O'Brien. 1998. Are Gulf landfalling hurricanes getting stronger? *Bull. Amer. Met. Soc.* 79, 1327–29.

Bryson, R. A., and G. J. Dittberner. 1976. A non-equilibrium model of hemispheric mean temperature.*J. Atm. Sci.* 33, 2094–2106.

Buchanan, J. B., and B. Tullock. 1962. *The Calculus of Consent*. Ann Arbor: University of Michigan Press, 361 pp.

Burnett, A. W. 1993. Size variations and long-wave circulation within the January Northern Hemisphere circumpolar vortex: 1946–89. *J. Climate* 6, 1914–20.

Cane, M. A., et al. 1997. Twentieth-century sea surface temperature trends. *Science* 275, 957–60.

Chan, J. C. L., and J. Shi. 1996. Long-term trends and interannual variability in tropical cyclone activity over the western North Pacific. *Geophys. Res. Let.* 23, 2765–2767.

Changnon, S. A. 1999. Impacts of 1997–1998 El Niño–generated weather in the United States, *Bull. Amer. Met. Soc.* 80, 1819–28.

Christy, J. R., and R. T. McNider. 1994. Satellite greenhouse signal. *Nature* 367, 325.

Christy, J. R., R. W. Spencer, and W. D. Braswell. 2000, MSU tropospheric temperatures: Dataset construction and radiosonde comparisons. *J. Atmos. and Oc. Tech* (in press).

Curtis, P. S., et al. 1994. Above- and below-ground response of *Populus grandidentata* to elevated atmospheric CO_2 and soil N availability. *Plant Soil* 165, 45–51.

Davis, R. E., and S. R. Benkovic. 1992. Climatological variations in the Northern Hemisphere circumpolar vortex in January. *Theor. App. Climatol.* 46, 63–73.

Davis, R. E., and R. Dolan. 1993. Nor'easters, *American Scientist* 81, 428–39.

Davis, R. E., P. C. Knappenberger, and A. Burnett. 1997. Relationships between surface temperatures and the 500 hPa circumpolar vortex in the Northern Hemisphere. *Proceedings 10th Conference of Applied Climatology*, 253–57.

Davis, R. E., et al. 1999. A climatology of snowfall-temperature relationships in Canada, *J. Geophys. Res.* 104, 11, 985–94.

Dlugokencky, E. J., et al. 1998. Continuing decline in the growth rate of atmospheric methane burden. *Nature* 393, 447–50.

DRI/McGraw-Hill Associates. 1998. Impacts of Kyoto. New York: McGraw-Hill Co. Inc.

Easterling, D. R., et al. 1997. Maximum and minimum temperature trends for the globe. *Science* 277, 364–67.

———. 1998. *Annual Energy Review*. Washington, D.C.: Energy Information Administration, U.S. Department of Energy.

Elsner, J. B., G. S. Lehmiller, and T. B. Kimberlain. 1996. Objective classification of Atlantic hurricanes. *J. Climate* 9, 2880–88.

Emanuel, K. A. 1986. An air-sea interaction theory for tropical cyclones. Part I: Steady-state maintenance. *Journal of the Atmospheric Sciences* 43, 585–604.

———. 1987. The dependence of hurricane intensity on climate. *Nature* 326, 483–85.

———. 1988a. The maximum intensity of hurricanes. *Journal of the Atmospheric Sciences* 45, 1143–56.

———. 1988b. Toward a general theory of hurricanes. *American Scientist* 76, 370–79.

———. 1995. Comments on "Global climate change and tropical cyclones": Part I. *Bull. Amer. Met. Soc.* 76, 2241–43.

Evans, J. L. 1993. Sensitivity of tropical cyclone intensity to sea-surface temperature, *J. Climate* 6, 1133–40

Evans, J. L., B. F. Ryan, and J. L. McGregor. 1994. A numerical exploration of the sensitivity of tropical cyclone rainfall intensity to sea surface temperatures. *J. Climate* 7, 616–23.

Fairbanks, R. G., et al. 1997. Evaluating climate indices and their geochemical proxies measured in corals. *Coral Reefs* 16, S93–S100.

Fan, S., et al. 1998. A large terrestrial sink in North America implied by atmospheric and oceanic carbon dioxide data and models, *Science* 282, 442–43.

Gore, A. 1992. *Earth in the Balance: Ecology and the Human Spirit.* New York: Plume, 407 pp.

Gray, W. M. 1984. Atlantic seasonal hurricane frequency. Part 1: El Niño and 30 mb quasi-biennial oscillation influences. *Monthly Weather Review* 112, 1649–68.

———. 1990. Strong association between West African rainfall and U. S. landfall of intense hurricanes. *Science* 249, 1251–56.

Gregory, J. M., and J. F. B. Mitchell. 1995. Simulation of daily variability of surface temperature and precipitation over Europe in the current and 2 × CO$_2$ climates using the UKMO climate model. *Quarterly Journal of the Royal Meteorological Society* 121, 1451–76.

Gregory, J. M., J. F. B. Mitchell, and A. J. Brady. 1997. Summer drought in northern mid-latitudes in a time-dependent CO$_2$ climate experiment. *J. Climate* 10, 662–86.

Grove, R. H. 1998. Global impact of the 1789–93 El Niño, *Nature* 393, 318–19.

Guilderson, T. M., and D. Schrag. 1998. Abrupt shift in subsurface temperature in the tropical Pacific associated with changes in El Niño. *Science* 281, 240–43.

Haarsma, R. J., J. F. B. Mitchell, and C. A. Senior. 1993. Tropical disturbances in a GCM. *Climate Dynamics* 8, 247–57.

Hansen, J. E. 1988. Testimony before the U.S. Congress on June 23, 1988.

Hansen, J. E., et al. 1981. Climate impact of increasing atmospheric carbon dioxide. *Science* 213, 957–66.

Hansen, J. E., et al. 1998. A common-sense climate index: Is our climate changing noticeably? *Proc. Nat. Acad. Sci.* 95, 4113–20.

Hansen, J. E., and A. A. Lacis. 1990. Sun and dust versus greenhouse gases: An assessment of their relative roles in global climate change. *Nature* 346, 713–18.

Hansen, J. E., M. Sato, and R. Ruedy. 1997. Radiative forcing and climate response. *J. Geophys. Res.* 102, 6831–34.

Hanson, K., G. A. Maul, and T. R. Karl. 1989. Are atmospheric "greenhouse" effects apparent in the climate record of the contiguous United States (1895–1987)? *Geophys. Res. Let.* 16, 49–52.

Harrison, D. E., and N. K. Larkin. 1997. Darwin sea level pressure, 1876–1996. Evidence for climate change? *Geophys. Res. Let.* 24, 1779–82.

Henderson-Sellers, A., and P. Robinson. 1986. *Contemporary Climatology.* Essex, England: Longman Scientific and Technical, 439 pp.

Henderson-Sellers, A., et al. 1998. Tropical cyclones and global climate change: A post-IPCC assessment. *Bull. Amer. Met. Soc.* 79, 19–38.

Hobbs, P. V., et al. 1997. Direct radiative forcing by smoke from biomass burning. *Science* 275, 1777–78.

Hobgood, J. S., and R. S. Cerveny. 1988. Ice-age hurricanes and tropical storms. *Nature* 333, 243–45.

Hoerling, M. P., and A. Kumar. 1997. Why do North American climate anomalies differ from one El Niño event to another? *Geophys. Res. Let.* 24, 1059–62.

Houghton, J. T. 1996. *London Times*, June 17.

Hughes, M. G., and D. A. Robinson. 1996. Historical snow cover variability in the Great Plains region of the USA: 1910 through to 1993, *Internat. J. Climatol.* 16, 1005–18.

Idso, K. E., and S. B. Idso. 1994. Plant responses to atmospheric CO_2 enrichment in the face of environmental constraints: A review of the past 10 years' research. *Agric. For. Meteorol.*, 69, 153–203.

Idso, S. B., R. C. Balling Jr., and R. S. Cerveny. 1990. Carbon dioxide and hurricanes: Implications of Northern Hemispheric warming for Atlantic/Caribbean storms. *Meteorol. Atmos. Phys.* 42, 259–63.

Intergovernmental Panel on Climate Change. 1990. *Climate Change: The IPCC Scientific Assessment.* J. T. Houghton, G. J. Jenkins, and J. J. Ephraums, eds. Cambridge: Cambridge University Press, 359 pp.

_____. 1992. *Climate Change 1992: The Supplementary Report to the IPCC Scientific Assessment.* Cambridge: Cambridge University Press, 200 pp.

_____. 1996. *Climate Change 1995: The Science of Climate Change: Contribution of Working Group I to the Second Assessment Report of the Intergovernmental Panel on Climate Change.* J. T. Houghton et al., eds. Cambridge: Cambridge University Press, 572 pp.

_____. 1999. *Special Report on Aviation and the Global Atmosphere.* Cambridge: Cambridge University Press, 373 pp.

Jones, P. D., et al. 1999. Surface air temperature and its changes over the past 150 years, *Rev. Geophys* 37, 173.

Kalkstein, L. S. 1991. A New Approach to Evaluate the Impact of Climate on Human Mortality, *Environmental Health Perspectives* 96, 145–50.

Kalkstein, L. S., and R. E. Davis. 1989. Weather and Human Mortality: An Evaluation of Demographic and Interregnal Responses in the United States. *Annals Assoc. Amer. Geog.* 79, 44–64.

Karl, T. R., and R. R. Heim Jr. 1990. Are droughts becoming more frequent or severe in the United States? *Geophys. Res. Let.* 17, 1921–24.

Karl, T. R., et al. 1995. Trends in U. S. climate during the twentieth century. *Consequences* 1, 3–12.

Karl, T. R., R. W. Knight, and N. Plummer. 1995. Trends in high-frequency climate variability in the twentieth century. *Nature* 377, 217–20.

Karl, T. R., et al. 1996. Indices of climate change for the United States. *Bull. Amer. Met. Soc.* 77, 279–92.

Karl, T. R., N. Nicholls, and J. Gregory. 1997. The coming climate. *Scientific American* 276, 79–83.

Kauppi, P. E., K. Mielikainen, and K. Kuusela. 1992. Biomass and carbon budget of European forests, *Science* 256, 70–74.

Kellogg, W. W., and Z. C. Zhao. 1988. Sensitivity of soil moisture to doubling of carbon dioxide in climate model experiments. Part I: North America. *J. Climate* 1, 348–66.

Kerr, J. B., and C. T. McElroy. 1993. Evidence for large upward trends of ultraviolet-B radiation linked to ozone depletion. *Science* 262, 1032–34.

Kerr, R. A. Model gets it right—without fudge factors. *Science* 276, 1041.

Kerr, R. A. Greenhouse forecasting still cloudy. *Science* 276, 1041–42.

King, J. S., et al. 1996. Growth and carbon accumulation in root systems of *Pinus taeda* and *Pinus ponderosa* seedlings as affected by varying CO_2, temperature and nitrogen. *Tree Physiol.* 16, 635–42.

Knappenberger, P. C., P. J. Michaels, and P. D. Schwartzman. 1996. Observed changes in the diurnal temperature and dewpoint cycles across the United States, *Geophys. Res. Let.* 23, 2637–40.

Knutson, T. R. and S. Manabe. 1994. Impact of increases in CO_2 on simulated ENSO-like phenomena. *Geophys. Res. Let.* 21, 2295–98.

Knutson, T. R., R. E. Tuleya, and Y. Kurihara. 1998. Simulated increase of hurricane intensities in a CO_2-warmed climate. *Science* 279, 1018–20.

Kuhn, T. S. 1962. *The Structure of Scientific Revolutions.* Chicago: Chicago University Press. 212 pp.

Laird, K. R., and S. C. Fritz. 1996. Greater drought intensity and frequency before A.D. 1200 in the Northern Great Plains, USA. *Nature* 384, 552–54.

Lamb, H. H. 1972. *Climate: Past, Present and Future: Volume I: Fundamentals and Climate Now.* Methuen & Company Ltd., London. 616 pp.

Lambert, S. J. 1996. Intense extratropical Northern Hemisphere winter cyclone events: 1899–1991. *J. Geophys. Res.* 101, 21, 319–25.

Landsea, C. W. 1993. A climatology of intense (or major) Atlantic hurricanes. *Monthly Weather Review* 121, 1703–13.

Landsea, C. W., et al. 1996. Downward trends in the frequency of intense Atlantic hurricanes during the past five decades. *Geophys. Res. Let.* 23, 1697–1700.

Lean, J., and D. Rind. 1998. Climate forcing by changing solar radiation. *J. Climate* 11, 3069–94.

Lenin Academy of Agricultural Sciences, USSR. 1949. *The Situation in Biological Science.* Moscow: Foreign Languages Publishing House, 631 pp.

Liang, X.-Z. and W.-C. Wang. 1998. The observed fingerprint of 1980–1997 ENSO evolution in the NCAR CSM equilibrium simulation. *Geophys. Res. Let.* 25, 1027–30.

Lighthill, J., et al. 1994. Global climate change and tropical cyclones. *Bull. Amer. Met. Soc.* 75, 2147–57.

Lins, H. F., and J. R. Slack. 1999. Streamflow trends in the United States, *Geophys. Res. Let.* 26, 227–30.

Long, S. P. 1991. Modification of the response of photosynthetic productivity to rising temperature by atmospheric CO_2 concentrations: Has its importance been underestimated? *Plant, Cell Envi.* 14, 729–39.

Manabe, S., and R. T. Wetherald. 1975. The effects of doubling the CO_2 concentraton on the climate of a general circulation model. *J. Atmos. Sci.* 32, 3–15.

Manabe, S., R. T. Wetherald, and R. J. Stouffer. 1981. Summer dryness due to an increase of atmospheric CO_2 concentration. *Climatic Change* 3, 347–86.

Manabe, S., and R. T. Wetherald. 1986. Reduction in summer soil wetness induced by an increase in atmospheric carbon dioxide. *Science* 232, 626–28.

Manabe, S., and R. T. Wetherald. 1987. Large-scale changes of soil wetness induced by an increase in atmospheric carbon dioxide. *Journal of the Atmospheric Sciences* 44, 1211–35.

Manabe, S., et al. 1991. Transient responses of a coupled ocean-atmosphere model to gradual changes of atmospheric CO_2: Part 1: Annual mean response. *J. Climate,* 4, 785–818.

McCabe, G. J., Jr., et al. 1990. Effects of climatic change on the Thornthwaite moisture index. *Water Resources Bulletin* 26, 633–43.

McGowan, J., et al. 1998. Climate-ocean variability and ecosystem response in the northeast Pacific. *Science* 281, 210–17.

McPhee, John. 1990 (reprint edition). *The Control of Nature.* New York: Noonday.

Meadows, D. H., and D. L., Meadows, eds. 1974. *The Limits to Growth: A Report for the Club of Rome's Project on the Predicament of Mankind.* New York: Universe, 205 pp.

Meadows, D., D. Meadows, and J. Rander, eds. 1993. *Beyond the Limits: Confronting Global Collapse, Envisioning a Sustainable Future,* (reprint edition). Chelsea Green Publishing Co., White River Junction, Vt. 320 pp.

Meehl, G. A., and W. M. Washington. 1996. El Niño-like climate change in a model increased atmospheric CO_2 concentrations, *Nature* 382, 56–60.

Michaels, P. J. 1992. *Sound and Fury: The Science and Politics of Global Warming,* Washington, D.C.: Cato Institute, 196 pp.

Michaels, P. J. 1999. A man ahead of his time: Discussions with a climate change pioneer: An interview with Reed A. Bryson, Ph.D. *State of the Climate Report: A World in Perspective.* Charlottesville, Va.: New Hope Environmental Services.

Michaels, P. J. 1998. Jungle fever: Tropical disease in the greenhouse: An interview with Paul Reiter, Ph.D. *State of the Climate Report: A World in Perspective.* Charlottesville, Va.: New Hope Environmental Services.

Michaels, P. J. 1998. *Los Angeles Times,* April 19.

Michaels, P. J., et al. 1990. Regional 500mb heights and U. S. 1000-500mb thickness prior to the radiosonde era. *Theor. Appl. Clim.* 42, 149–54.

Michaels, P. J. and D. E. Stooksbury. 1992. Global warming: A reduced threat. *Bull. Amer. Met. Soc.* 73, 1536–77.

———. 1993. Reply comment to 1992 article. *Bull. Amer. Met. Soc.* 74, 856–57.

Michaels, P. J., et al. 1994. General circulation models: Testing the forecast. *Technology* 331A, 123–33.

Michaels, P. J., and H. Lins. 1994. Increasing U.S. streamflow linked to greenhouse forcing, *EOS Transactions of the American Geophysical Union* 76, 281–85.

Michaels, P. J., S. F. Singer, and P. C. Knappenberger. 1994. Analyzing Ultraviolet-B radiation: Is there a trend? *Science* 264, 1341–42.

Michaels, P. J., and P. C. Knappenberger. 1996. Human effect on Global Climate? *Nature* 384, 522–23.

Michaels, P. J., et al. 1998. Analysis of trends in the variability of daily and monthly historical temperature measurements. *Clim. Res.* 10, 27–33.

Mitchell, J. F. B., and D. A. Warrilow. 1987. Summer dryness in northern mid-latitudes due to increased CO_2. *Nature* 330, 238–40.

Mitchell, J. F. B., et al. 1995. Climate response to increasing levels of greenhouse gases and sulfate aerosols. *Nature* 376, 501–4.

Mitchell, J. F. B, et al. 1995. On surface temperature, greenhouse gases, and aerosols: models and observations, *J. Climate* 8, 2364–86.

Mitchell, J. F. B., and T. C. Johns. 1997. On modifications of global warming by sulfate aerosols. *J. Climate* 10, 245–66.

Moore, T. G. 1998. *Climate of Fear: Why We Shouldn't Worry about Global Warming,* Washington, D.C.: Cato Institute, 175 pp.

Mora, C. I., S. G. Driese, and L. A. Colarusso. 1996. Middle to late Paleozoic atmospheric CO_2 levels from soil carbonate and organic matter. *Science* 271, 1105–07.

Murphy, J. M. and J. F. B. Mitchell. 1995. Transient response of the Hadley Centre coupled ocean-atmosphere model to increasing carbon dioxide, Part II: Spatial and temporal structure of response. *J. Climate* 8, 57–80.

Myhre, G. 1998. New estimates of radiative forcing due to well-mixed greenhouse gases *Geophys. Res. Let.* 25, 2715–18.

Myneni, R. B. et al. 1997. Increased plant growth in the northern high latitudes, from 1981 to 1991, *Nature* 386, 698–702.

———. *Newsweek,* 1996. January 22.

National Research Council. 2000. *Reconciling Observations of Global Temperature Change.* National Academy Press, Washington, D.C. 85 pp.

Nicholls, N. 1984. The Southern Oscillation, sea surface temperature, and interannual fluctuations in Australian tropical cyclone activity, *International Journal of Climatology* 4, 661–670.

Nicholls, N., C. Landsea, and J. Gill. 1998. Recent trends in Australian region tropical cyclone activity. *Meteorology and Atmospheric Physics* 65, 197–205.

Niskanen, W. A. 1997. R&D and Economic Growth—Cautionary Thoughts. In Barfield, C. E., Ed., *Science for the 21st Century,* Washington D.C.: AEI Press, pp. 81–94.

Ohmura, A., M. Wild, and L. Bengtsson. 1996. A possible change in mass balance of Greenland and Antarctic ice sheets in the coming century. *J. Climate* 9, 2124–35.

Oort, A. H., et al. 1989. Historical trends in surface temperature over the oceans based on the COADS. *Climate Dynamics* 2, 89.

Palmer, W. C. 1965. Meteorological drought. *Research Paper 45.* Washington, D.C.: U.S. Weather Bureau.

Parker, D. E., et al. 1994. Interdecadal changes of surface temperature since the late nineteenth century. *J. Geophys. Res.* 99, 14, 373–99.

Parker, D. E., T. P. Legg, and C. K. Folland. 1992. A new daily central England temperature series, 1772–1991. *International Journal of Climatology* 12, 317–42.

Peterson, T. C., V. S. Golubev, and P. Ya. Groisman. 1995. Evaporation losing its strength, *Nature* 377, 687–88.

Pielke, R. A., Sr., et al. 1998. Interactions between the atmosphere and terrestrial ecosystems: Influence on weather and climate. *Global Change Biology* 4, 461–75.

Poorter, H. 1993. Interspecific variation in the growth response of plants to an elevated and ambient CO_2 concentration. *Vegetation* 104/105, 77–97.

Prior, S. A., et al. 1995. Free-air carbon dioxide enrichment of cotton: Root morphological characteristics. *Journal of Environmental Quality* 24, 678–83.

Ramanathan, V., et al. 1988. Cloud-radiative forcing and climate: Results from the Earth Radiation Budget Experiment. *Science* 243, 53–67.

Reiter, P. 1996. Global warming and mosquito-borne disease in the USA. *The Lancet* 348, 662.

Rind, D., et al. 1990. Potential evapotranspiration and the likelihood of future drought. *J. Geophys. Res.* 95, 9983–10004.

Rind, D., C. Rosenzweig, and R. Goldberg. 1992. Modeling the hydrological cycle in assessments of climate change. *Nature* 358, 119–22.

Royer, J.-F., et al. 1998. A GCM study of the impact of greenhouse gas increase on the frequency of occurrence of tropical cyclones. *Climatic Change* 38, 307–43.

Ryan, B. F., I. G. Watterson, and J. L. Evans. 1992. Tropical cyclone frequencies inferred from Gray's yearly genesis parameter: Validation of GCM tropical climates. *Geophys. Res. Let.* 19, 1831–34.

Sandweiss, D. H., et al. 1996. Geoarchaeological evidence from Peru for a 5,000-year B. P. onset of El Niño. *Science* 273, 1531–33.

Santer, B. D., et al. 1996. A search for human influences on the thermal structure of the atmosphere. *Nature* 382, 36–45.

Santer, B. D., et al. 2000. Interpreting differential temperature trends at the surface and in the lower troposphere. *Science,* 287, 1227–1232.

Schell, J. 1989, *Discover,* October.

Schiesser, H. H. et al. 1997. Winter storms in Switzerland north of the Alps, 1864/ 1865–1993/1994. *Theor App. Climatol.* 58, 1–19.

Schneider, S. H., and R. Chen. 1980. Carbon dioxide warming and coastal flooding: Physical factors and climate impact. *Annual Review of Energy* 5, 107–35.

Schwartzman, P. D., P. J. Michaels, and P. C. Knappenberger. 1998. Observed changes in the diurnal dewpoint cycles across North America, *Geophys. Res. Let.* 25, 2265–68.

Serreze, M. C., et al. 1997. Icelandic Low cyclone activity: Climatological features, linkages with the NAO, and relationships with recent changes in the Northern Hemisphere circulation, *J. Climate* 20, 453–64.

Sietz, F. 1996. *Wall Street Journal*, June 12.

Simard, S. W., et al. 1997. Net transfer of carbon between ectomycorrhizal tree species in the field. *Nature,* 388, 579–82.

Soulé, P. T. 1990. Spatial patterns of multiple drought types in the contiguous United States: A seasonal comparison. *Clim. Res.* 1, 13–21.

Soulé, P. T. 1992. Spatial patterns of frequency and duration for persistent near-normal climatic events in the contiguous United States. *Clim. Res.* 2, 81–89.

Soulé, P. T. 1993. Hydrologic drought in the contiguous United States, 1900–1989: Spatial patterns and multiple comparison of means. *Geophys. Res. Let.* 20, 2367–70.

Soulé, P. T., and Z. Y. Yin. 1995. Short- to long-term trends in hydrologic drought conditions in the contiguous United States. *Clim. Res.* 5, 149–57.

Spencer, R., and J. E. Christy. 1991. Precise monitoring of global temperature trends from satellites. *Science* 247, 1558–62 (and updates).

Stevens, B. 1995. *New York Times*, September 10.

Sun, de-Zheng. 1997. El Niño: A coupled response to radiative heating. *Geophys. Res. Let.* 24, 2031–34.

Taylor, K., and J. E. Penner. 1994. Climate system response to aerosols and greenhouse gases: A model study. *Nature* 369, 734–37.

Teeri, J. A., and L. G. Stowe. 1976. Climatic patterns and the distribution of C4 grasses in North America. *Oecologia* 23, 1–12.

Tegen, I., A. A. Lacis, and I. Fung. 1996. The influence on climate forcing of mineral aerosols from disturbed soils. *Nature* 380, 419–22.

Thomson, D. J. 1995. The seasons, global temperature, and precession, *Science* 268, 59–68.

Trenberth, K. E. 1998. El Niño and global warming: What's the connection? *UCAR Quarterly* 24, Winter 1997.

Trenberth, K.E. 1998. Testimony to the Subcommittee on Science, Technology, and Space, U.S. Senate, February 26, 1992.

Trenberth, K. E., G. W. Branstator, and P. A. Arkin. 1988. Origins of the 1988 North American drought, *Science* 242, 1640–45.

Trenberth, K. E., and G. W. Branstator. 1992. Issues in establishing causes of the 1988 drought over North America. *J. Climate* 5, 159–72.

Trenberth, K. E., and C. J. Guillemot. 1996. Physical processes involved in the 1988 drought and 1993 floods in North America. *J. Climate* 9, 1288–98.

Trenberth, K. E., and T. J. Hoar. 1996. The 1990–1995 El Niño-Southern Oscillation event: Longest on record. *Geophys. Res. Let.* 23, 57–60.

Tsonis, A. A., P. J. Roebber, and J. B. Elsner. 1998. A characteristic time scale in the global temperature record, *Geophys. Res. Let.* 25, 2821–23.

———. 1985. State of the Art Series. Washington, D.C.: U.S. Department of Energy.

Vinnikov, K. Ya., et al. 1996. Scales of temporal and spatial variability of midlatitude soil moisture. *Geophys. Res. Let.*

Vinnikov, K. Ya., et al. 1996. Vertical patterns of free and forced climate variations. *Geophys. Res. Let.* Vol. 23 , No. 14 , p. 1801–1804.

Vital Statistics of the United States, U.S. Department of Commerce. Various years. *Statistical Abstract of the United States.* Washington, D.C.: Government Printing Office.

Volin, J. C., et al. 1998. Elevated carbon dioxide ameliorates the effects of ozone on photosynthesis and growth: Species respond similarly regardless of photosynthetic pathway or plant functional group. *New Phytologist* 138, 315–25.

Wagner, F., et al. 1999. Century-scale shifts in early Holocene atmospheric CO_2 concentration. Science 284, 1971–73.

WEFA Inc. 1998. *Global Warming: The high costs of the Kyoto Protocol.*

Wentz, F. J., and M. Schabel. 1998. Effects of orbital decay on satellite-derived lower-tropospheric temperature trends. *Nature* 384, 661–64.

Wetherald, R. T., and S. Manabe. 1995. The mechanisms of summer dryness induced by greenhouse warming. *J. Climate* 8, 3096–3108.

Whitney, L. D., and J. S. Hobgood. 1997. The relationship between sea surface temperatures and maximum intensities of tropical cyclones in the eastern North Pacific Ocean. *J. Climate,* 10, 2921–30.

Wigley, T. M. L. 1987. Relative contributions of different trace gases to the greenhouse effect. *Climate Monitor* 16, 14–28.

———. 1998. Stumbling at the start of a marathon. *Earth,* April, 54–55.

———. 1998. The Kyoto Protocol: CO_2, CH4, and climate implications. *Geophys. Res. Let.,* 25, 2285–88.

———. NCAR press release, April 14, 1999.

———. 1999. The science of climate change. Pew Center on Global Climate Change, The Pew Foundation.

Wittwer, S. H. 1995. *Food, Climate, and Carbon Dioxide.* Boca Raton, Fla.: CRC Press, 236 pp.

———. *World Survey of Climatology* series (various eds.), New York: Elsevier.

Zelilch, I. 1992. Control of plant productivity by regulation of photorespiration. *Biological Science* 42, 510–16.

Zhang, Y. and W. C. Wang. 1995. Model-simulated Northern winter cyclone and anticyclone activity under a greenhouse warming scenario. *J. Climate* 10, 1616–34.

Index

A Consortium for the Analysis of Climate Impact Assessments (ACACIA), 71–72
Aerosols. *See* Sulfate aerosols
Albedo, or reflectivity
  changes in earth's, 60
  defined, 25
  of earth's surface, 59
Angel, J. R., 151
Angell, J. K., 82, 85, 97, 99, 150
Anticyclones
  high-pressure systems as, 42
  polar, 40
  Siberian and North American, 43
  subtropical, 40
  tropical and polar, 42–43
Atmosphere
  carbon dioxide in, 26
  changing composition of, 57
  with higher levels of CO_2, 178–79
  levels of methane in, 31–32
  molecules intercepting earth's radiation, 25
  response to El Niño conditions, 43–45
  upward and downward motion in, 41
  water vapor in, 25–26
  See also Circulation system, atmospheric; Temperature, atmospheric

Balling, R. C., Jr., 112, 127, 128, 143
Beersma, J. J., 149
Begley, Sharon, 152
Bengtsson, L., 141–42
Benkovic, Stephanie, 150
Bentley, Charles, 163
Blizzards, 151–52
Bodansky, Daniel, 14, 17
Bonan, Gordon, 11
Bottcher, Frits, 13–14
Bove, M. C., 145
Brown, George, 191
Bryson, Reid, 48, 65

Buchanan, James
  The Calculus of Consent, 5, 10, 192–95
Burnett, Aaron, 150
Byrd-Hagel Resolution, 1998 (S. Res. 98), 202

Calculus of Consent, The (Buchanan and Tullock), 5, 10, 192–95
Callendar, Bruce, 19
Cane, M. A., 102
Carbon dioxide
  climate changes related to (1980s–90s), 116–18
  direct effects on human health, 176
  fertilization effect on plants, 178
  from fires and volcanoes, 26
  human contributions of, 26–27
  IPCC scenarios for future effective concentration, 37–39
  link to drought conditions, 116–24
  link to hurricane formation, 141–42
  models to evaluate impact on soil moisture, 117
  nonhuman events changing greenhouse effect, 26
  photosynthesis response of plants to, 179–81, 183–84
  radiation in the troposphere from, 28
  relation to crop yield levels, 185–90
  rise in concentrations, 31
  as source of greenhouse effect, 26–27
Carbon dioxide emissions
  effect of, 4
  increased, 209–10
Center for International Earth Science Information Network (CIESIN), 72
CFCs. *See* Chlorofluorocarbons (CFCs)
Chan, J. C. L., 144
Changnon, Stanley, 50
Charlson, Robert, 69
China
  as source of sulfate aerosols, 73
Chlorofluorocarbons (CFCs)
  as cause of ozone depletion, 93–96
  omitted from FCCC, 200

About the Authors

Patrick J. Michaels is a research professor of environmental sciences at the University of Virginia and visiting scientist with the Marshall Institute in Washington, D.C. He is a past president of the American Association of State Climatologists and was program chair for the Committee on Applied Climatology of the American Meteorological Society. He holds A.B. and S.M. degrees in biological sciences and plant ecology from the University of Chicago, and he received a Ph.D. in ecological climatology from the University of Wisconsin-Madison in 1979. Michaels is a contributing author and reviewer of the United Nations Intergovernmental Panel on Climate Change (IPCC). He has published more than 200 scientific, technical, and popular articles on climate and its impact on man and is the author of *Sound and Fury: The Science and Politics of Global Warming*, published in 1992 by the Cato Institute, where he is a senior fellow in environmental studies. According to *Nature* magazine, Pat Michaels may be the most popular lecturer in the nation on the subject of global warming.

Robert C. Balling, Jr., is Professor of Geography and Director of the Office of Climatology at Arizona State University. He is the author of over 100 papers in the leading scientific journals of climate and atmospheric sciences. He is a world authority on drought and climate change and was selected by the United Nations as author of their definitive book, *Interactions of Desertification and Climate*. Dr. Balling is internationally recognized as one of the world's premier lecturers on the subject of human-induced climate change.

Cato Institute

Founded in 1977, the Cato Institute is a public policy research foundation dedicated to broadening the parameters of policy debate to allow consideration of more options that are consistent with the traditional American principles of limited government, individual liberty, and peace. To that end, the Institute strives to achieve greater involvement of the intelligent, concerned lay public in questions of policy and the proper role of government.

The Institute is named for *Cato's Letters,* libertarian pamphlets that were widely read in the American Colonies in the early 18th century and played a major role in laying the philosophical foundation for the American Revolution.

Despite the achievement of the nation's Founders, today virtually no aspect of life is free from government encroachment. A pervasive intolerance for individual rights is shown by government's arbitrary intrusions into private economic transactions and its disregard for civil liberties.

To counter that trend, the Cato Institute undertakes an extensive publications program that addresses the complete spectrum of policy issues. Books, monographs, and shorter studies are commissioned to examine the federal budget, Social Security, regulation, military spending, international trade, and myriad other issues. Major policy conferences are held throughout the year, from which papers are published thrice yearly in the *Cato Journal.* The Institute also publishes the quarterly magazine *Regulation.*

In order to maintain its independence, the Cato Institute accepts no government funding. Contributions are received from foundations, corporations, and individuals, and other revenue is generated from the sale of publications. The Institute is a nonprofit, tax-exempt, educational foundation under Section 501(c)3 of the Internal Revenue Code.

CATO INSTITUTE
1000 Massachusetts Ave., N.W.
Washington, D.C. 20001